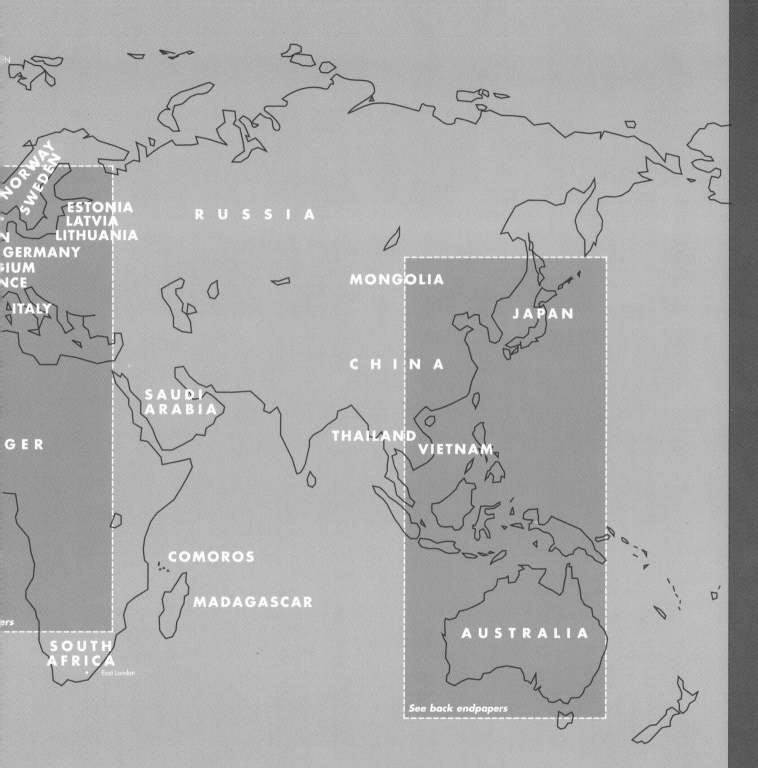

NORWAY
SWEDEN
ESTONIA
LATVIA
LITHUANIA
GERMANY
GIUM
NCE
ITALY

R U S S I A

MONGOLIA

JAPAN

C H I N A

SAUDI
ARABIA

THAILAND
VIETNAM

GER

COMOROS

MADAGASCAR

AUSTRALIA

ers

SOUTH
AFRICA
East London

See back endpapers

F O S S I L   S I T E S
pal regions and territories covered

THE RISE OF FISHES
500 million years of evolution

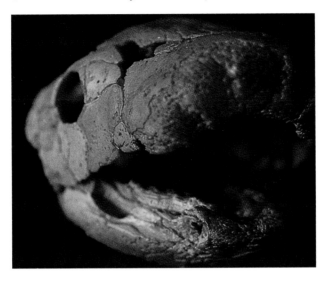

THE RISE

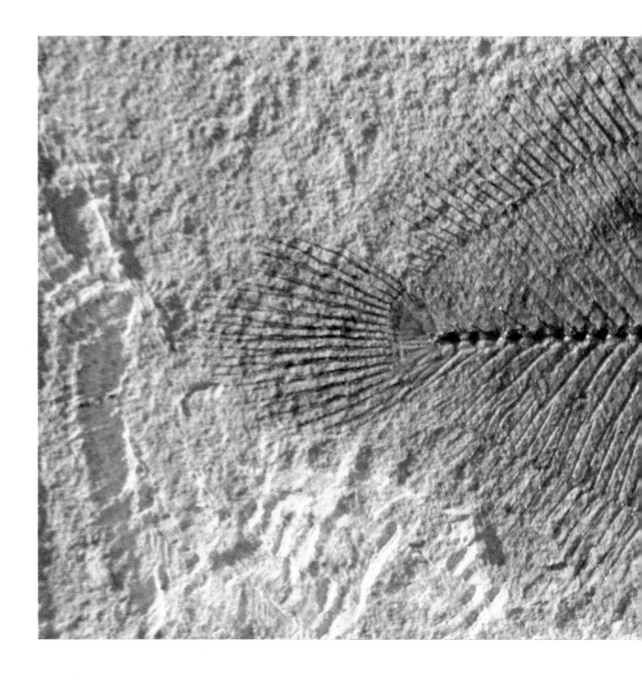

THE JOHNS HOPKINS UNIVERSITY PRESS
BALTIMORE AND LONDON

OF FISHES

500 million years of evolution

JOHN A. LONG

CURATOR OF VERTEBRATE PALAEONTOLOGY,
WESTERN AUSTRALIAN MUSEUM,
PERTH, WESTERN AUSTRALIA

Published in Australia by
University of New South Wales Press

Published in the United States of America by
The Johns Hopkins University Press,
2715 North Charles Street
Baltimore, Maryland 21218-4319
The Johns Hopkins Press Ltd., London

Library of Congress Cataloging-in-Publication Data:

Long, John A., 1957– .
The rise of fishes : 500 million years of evolution / John A. Long
p. cm.

Includes bibliographical references and index.
ISBN 0-8018-4992-6
1.Fishes — Evolution. I.Title.
QL618.2.L66 1994   94-24692
597'.038—dc20   CIP

A catalog record for this book is available from the British Library.
Designed by Di Quick
Printed by Kyodo Printing, Singapore

Previous pages: Eobothus, a sole from Eocene of Monte Bolca. *Dr Lorenzo Sorbini*
Half-title page: Chirodipterus australis, a Devonian lungfish from the Gogo
 formation of Western Australia. *John A. Long*

Contents

Acknowledgements

This book is the end result of years of working with many fine people in the field of fish evolution. As a student my thesis work was guided and encouraged by Dr Pat Vickers-Rich and Dr Jim Warren of Monash University, who first directed me into the fascinating study of fossil fishes. Since then I have received generous support and help from my Australian colleagues, particularly Dr Alex Ritchie, Dr Gavin Young, Professor Ken Campbell, Dr Dick Barwick and Dr Sue Turner. I would like to thank these people for permission to use freely of their work and, in some cases, for provision of photographs and artwork reproduced throughout this book.

Several colleagues from around the world were enthusiastic about the book and discussed aspects of their new research or assisted with obtaining illustrations, particularly: Dr Mark V. Wilson (Canada), Dr Marius Arsenault (Parc de Miguashua, Quebec), David Ward (Orpington, UK), Dr Mike Coates and Dr Jenny Clack (Cambridge University Museum, UK), Dr Karl Frickhinger (Germany), Dr Philippe Janvier, Dr Daniel Goujet and Dr Hervé Lelievre (Natural History Museum, Paris, France), Dr Richard Cloutier and Dr Alain Blieck (Lille University, France); Dr Oleg Lebedev (Moscow, Russia); Dr Pierre Yves-Gagnier (McGill University, Canada), Dr Hans Bjerring, Professor Erik Jarvik and Professor Tor Ørvig (Swedish Museum of Natural History, Stockholm); Dr Per Ahlberg (Oxford University, UK), Professor Hans-Peter Schultze (Humboldt Museum, Berlin, Germany); Dr Mike Williams (Cleveland Museum, Ohio, USA), Dr Dick Lund (Adelphi University, New York); Dr Nigel Trewin (Aberdeen University, Scotland); Dr Lance Grande (Field Museum, Chicago, USA), Dr John Maisey (American Museum of Natural History, New York); Dr Peter Forey, Dr Colin Patterson, Dr Roger Miles and Professor Brian Gardiner (Natural History Museum, London); Dr Dick Aldridge and Dr Derek Briggs (Leicester University, UK), Dr Moya Smith (Guy's Dental Hospital, London). Also thanks to Stan Wood for helpful discussion of Scottish fossil fish sites and his hospitality during my stay in Edinburgh.

The following people are thanked for contributing photographs for use in this book or helping arrange photographs: Mr Douglas Elford, Ms Kristine Brimmell, Ms Loisette Marsh and Mr Clay Bryce (Western Australian Museum, Perth); Dr Ralph Molnar (Queensland Museum); Ron and Valerie Taylor (Sydney); Dean Lee (Underwater World, Perth); Professor Ian Potter (Murdoch University, Perth); Dr Neil Clarke (Hunterian Museum, Glasgow) and T. Nishiinoue (Kagoshima University, Japan). Thanks also to Dr Gerry Allen and Dr Barry Hutchinson (Western Australian Museum) for their many helpful discussions about fishes in general.

I thank Rex Parry from University of New South Wales Press who showed faith in my proposal and got the project off the ground, then steered it throughout its course. Rex gave me much encouragement and helpful editorial advice to greatly improve the final version of this book. I would also like to thank the publisher's whole team for the great job done on the production of the book. Thanks also to Sue, Therese and Donna of Professional Presentations, West Perth, for their expert advice on production of the computer graphics.

I would like to thank my colleagues at the Western Australian Museum, Dr Ken McNamara, Dr Alex Bevan, Dr Alex Baynes and Ms Kristine Brimmell for their enthusiasm and continued support of my research. Ken is especially thanked for critically reading and commenting on the first draft of this book. I also thank the Director and Trustees of the Western Australian Museum for their support of publishing popular science works as a vital part of the museum's role in providing information to all levels of the community.

Preface

The rise of fishes from the waters of the archaic seas and rivers onto land signifies one of the greatest events in the evolution of life on Earth. If not for the evolution of four-legged land animals from fish, we humans, belonging to a small group of warm-blooded land animals called "mammals", would never have had an opportunity to evolve. For we are merely highly advanced fishes. Yet representation of this event is often given short treatment in the story of the evolution of backboned animals (vertebrates), as such accounts tend to focus on the mainstream evolution of land animals—from amphibians to reptiles, birds and mammals, and ultimately to us.

Fishes play an integral role in human survival. They are one of our important food sources and part of the oceanic food chain on which the planet's balanced ecosystem is totally dependent. This book tells the remarkable story of the early evolution of fishes, culminating in the ultimate transformation from water-dwelling fishes into air-breathing animals with limbs.

The evolution of fishes spans a fascinating 500 million years, of which the first 150 million years was a time when fishes lived on the Earth as the pinnacle of evolutionary achievement. Many strange families of fishes evolved and just as mysteriously disappeared from this planet more than 120 million years before the dawn of the first dinosaurs.

Fishes are the most diverse group of vertebrates, with more than 23,000 known species inhabiting our seas, rivers and lakes. People have a close dependence on living fishes. They are not only a major food resource, but they also form the basis of our largest recreational sport. We also like to keep them as pets. Yet few of us have any idea of where and how the first fishes originated or in what way their complex evolution has played a key role in our own ancestry. This book takes us on a long and fascinating journey through the Earth's distant ages to unravel the amazing story of the evolution of fishes. We shall go back to the dawn of higher life on Earth, to a point in time some 500 million years ago.

Fishes evolved as the first creatures to have a skeleton, making them the ancestors of all vertebrates—amphibians, reptiles, birds and mammals. Many bizarre and unusual groups of fishes evolved early in their history but did not make the grade. It is their story that this book aims to tell, largely from the perspective of Gondwana, the great southern supercontinent. The reason for this is simply that much of fish evolution was centred in the waters of the ancient supercontinent of Gondwana, and many of the world's finest fish fossils occur in some of today's remnants of Gondwana—South America, Africa, India, Antarctica and Australia. The overall aim of the book is to show the complete story of fish evolution, incorporating many of the world's best examples.

The study of fossil fish is interesting not just because we want to know more about our distant origins and the changing forms of life on Earth. Fish fossils are also useful in assessing the ages of rocks in the Palaeozoic Era, and help us to solve problems of geological correlation and plate tectonic movements. In some countries the mining of fish fossils is a several-million-dollar-a-year industry, not only through their sale and trade but also through the growing industry of scientific tourism. Whatever their use or their economic applications, fish fossils remain intrinsically beautiful and often mysterious objects which have long captured the imagination of people.

Finally, this book is intended not only as a simple guide to the many intriguing types of fish fossils and their unique evolutionary histories. It is also a visual celebration of their essential form and beauty when seen as once-living creatures, now known to the world only as crystallised remains in stone.

Introduction

The story of fishes through time is also the story of changing continents and climates, devastating mass extinctions and changing faunas and floras. The complete picture can be grasped only from a multifaceted approach, and this is why there are several sections here giving extra information for non-palaeontologists. A glossary of complex or scientific terms is included in the back for ready reference.

LIFE HISTORIES IN STONE

Fossils are the remains of former life on Earth. These are usually organic materials that have been buried by natural processes and altered in some chemical way before being exposed again at the surfaces of the Earth, often because of the dynamic processes of uplift and erosion. They can take the form of skeletons, shells, impressions or trackways and in some instances can be the whole organism completely preserved, such as fossil insects entrapped in amber. The word fossil comes from a Latin word *fossilis*, meaning "dug up" (past tense of the verb *fodere*, to dig).

The burial of organic material within layers of sediment that becomes rock generally results in alteration or mineralisation of the original organic material. Bones, shells, wood and other porous hard parts of animals and plants can become replaced by different minerals. Calcite and aragonite, both forms of calcium carbonate, commonly replace shell material. Bone can be replaced by opal (cryptocrystalline silica) in some cases, although in most fossil fishes the bones retain their original mineral composition but are added to by secondary minerals seeping into the pore spaces in the bones, changing the colour of fossil bones and teeth from white (or opaque) to brown or black. Bone is made up of the mineral hydroxylapatite (calcium

phosphate hydroxide). In favourable conditions, with an abundant source of calcite, the calcium can replace the phosphorus as water is removed, thus strengthening and preserving the bone; this is called calcification. In cases where groundwaters replace the bone with silica or enrich the bone with silica minerals, it is termed silicification.

Fish fossils are most commonly found as skeletons compressed within the layers of rock that built up after drifting muds or sands covered the decayed carcasses. By splitting the finely layered shale or mudstone from certain sites, whole fish skeletons millions of years old can be found. Such fossils provide us with an extraordinary amount of information about the once-living organism when we compare the skeletal anatomy of the fossil fish with that of living forms. In rare cases some fish fossils are preserved as uncrushed, three-dimensionally perfect whole skeletons. These fossils occur in limestone rocks where the formation of the carbonate rock around the uncrushed dead fish was very rapid. Today, using chemical preparation techniques (such as applying weak acids), we can physically free the bones from the rock without damaging the fossil. Examples of such extraordinary preservation are seen throughout this book in the 370-million-year-old fishes from the Gogo Formation of Western Australia, and the 390-million-year-old fishes from the Taemas and Wee Jasper regions of New South Wales in southeastern Australia.

Douglas Elford, Western Australian Museum

At the other extreme, some fishes are known only from one small and rather insignificant fossil remain: their ear-stones or otoliths. Such fossils are made of material denser than bone and are preserved in deep oceanic sediments. They can be sieved out of sediment samples and are often found in

◀ Fossil shells *Trigonia moorei*, Jurassic Period, Western Australia

▲ Living stromatolites, Shark Bay, Western Australia, are representative of the oldest surviving forms of life on Earth.

great abundance. The complex shapes of the ridges and grooves on these ear-stones allows us to readily identify the exact species of fish they once belonged to. However, this sort of analysis really only applies to fossil remains of groups of fishes that are still alive today, by referring to the ear-stones described in modern fishes. Yet fish otoliths actually date back as far as the Devonian Period, more than 360 million years ago. These have a simple form and structure, unlike those of modern fishes, and are generally found intact only within the head of the fish fossil.

FOSSILS AND ROCKS

Throughout this book there are many references to rock types that enclose the fish fossils. All fossils are preserved in sedimentary rocks—rocks formed by the cementation and compression of many grains of sediment. Sedimentary rocks that commonly contain fossils are either clastic or carbonate rocks. Clastic rock is made up of separate grains called "clasts". Fine-grained particles, such as mud or silt, under burial, form mudstones, siltstones or flat-layered fissile rocks like shale, whereas coarser-grained rocks form sandstones or conglomerates.

John A. Long

◄ Sedimentary rocks form in basins or subsiding areas of the Earth's crust. They can be coarse-grained grits or conglomerates, or fine-grained rocks like mudstones and shale. This is the Beacon Sandstone at Sponsor's Bluff, south Victoria Land, Antarctica.

Carbonate rocks are bound together by calcium carbonate. Limestones are carbonate rocks formed from calcium carbonate yet may also include a proportion of mud or silt or sand. Limestones are quite often made up of the accumulated remains of many small organisms that have calcitic shells, such as clams, or microscopic fossils such as foraminifers.

The term "sandstone" applies to rocks formed principally of sand grains, and these are defined only by the size of grains, being larger than 0.2 mm and smaller than 2 mm in diameter. Grains larger than 2 mm can form "coarse-grained" sediments like grit and conglomerate; these are formed in high-energy environments where fossils are often broken up before final burial. Sandstones can be made up of grains of various minerals, but quartz and feldspar are the commonest varieties. A good rule of thumb is that the finer the sediment particle size, the quieter the environment in which the rock formed, and the better

the chance of a complete, well-preserved fossil skeleton. Very fine-grained muds accumulate in deep still waters, whereas active high-energy environments such as riverbeds and tidal zones tend to be dominated by sandstones or coarser-grained rocks. Very coarse sedimentary rocks, such as conglomerates, generally lack good fossils because they were deposited in turbulent conditions where delicate organic remains were broken and not preserved.

Many fish fossils of the Devonian Period are preserved in "red beds", a common name for red-coloured sandstones, mudstones and shales that represent river, lake and estuarine deposits. Many famous sites in Scotland occur in such rocks, which were first named the "Old Red Sandstone". This term now applies to a widespread belt of Devonian river and shallow marine rocks occurring throughout the ancient supercontinent of Laurasia (sometimes called "The Old Red Continent"). "Redbeds" get their coloration from the oxidation of iron–rich sediments in warm to tropical dry climates. In most cases they may include green shales and mudstones as well as ancient soil deposits. Fish fossils may accumulate in ancient channel and lake deposits within these rock sequences.

▲ A fish has died and sinks to the bottom of the river, lake or sea, soon to be buried by incoming sediment settling out of the water.

▲ Millions of years later the sediment (mud) containing the fish skeleton has been compressed by the weight of many layers of sediment above it. It has been turned to mudstone. The bones are enriched by new minerals and become fossilised.

▲ Uplift of the earth's crust has exposed these sedimentary rock layers in a cliff face. Erosion eventually exposes the fossil skeleton.

Douglas Elford, Western Australian Museum

John A. Long

Fossils are the remains of former life on this planet such as bones, shells, leaves or wood. They also include impressions of where animals once walked. Fossils are generally preserved by replacement of the original material with other minerals or, in the case of bone, may be strengthened by the addition of calcium minerals replacing the phosphorus component.

▲ Ammonite *Newmarracaroceras*, Jurassic Period, Western Australia.

◀ Dinosaur footprint from the Cretaceous Period, Western Australia.

▼ Brachiopods, Permian Period, Western Australia.

John A. Long

A POTTED EVOLUTIONARY HISTORY OF LIFE

Fossils tell a remarkable story of the transformation of life through time—evolution—and understanding evolution involves grasping the concepts of geological time. It is the vast span of the Earth's age, its "deep time", that underwrites the credibility of complex series of changes in species mutating to form new species. The age of the Earth has been subdivided into blocks of time, based on major events in the development of life.

LARVAE AND EVOLUTION

The origin of the first fishes from invertebrates appears to be intimately linked with the major changes in shape exhibited by certain invertebrates during their development. The juvenile phases of fishes and invertebrates are often quite morphologically different from their adult forms. As adults sea-squirts (tunicates) are sessile filter-feeding organisms living attached to the sea floor, although their larvae are free-swimming forms with long muscular tails and a well-developed notochord, dorsal hollow nerve cord, gill slits and strengthened mineralised rods supporting the tail. They share the latter features with all chordates (animals with a notochord), and some of these features occur in enigmatic protochordates. Usually such major changes in morphology within the lifespan of an organism are the result of a change in habitat as, for example, in the tunicate larvae moving from the open waters to the seabed.

Other invertebrate groups similarly have mobile larvae and grow into less-agile adult forms (for example, echinoderms), and theories have been advocated that they too may have given rise to the first fishes. The first fishes may well have evolved from retention of juvenile stages in such free-swimming larvae by precocious sexual maturation, enabling some free-swimming larvae to breed.

The larvae of the marine lamprey *Petromyzon marinus* are called ammocoetes. The larvae live in burrows in sandy river bottoms for three years or more before undergoing metamorphosis. The young lampreys then migrate down the river to the sea, where they remain for another three to four years before becoming sexually mature. The ammocoete is a good example of primitive vertebrate organisation and is not dissimilar from the generalised tunicate larvae or the cephalochordate *Branchiostoma*.

The degree of changes that a single individual fish goes through in its life history may also reflect the large-scale evolutionary trends inherent within its group. For example, the living Queensland lungfish *Neoceratodus* has a body form similar to that of its larvae, with little change to the shape and size of the fins during growth. Lungfish of the Devonian Period had similar larvae (where known) but developed into adults with widely differing morphology—separate dorsal and anal fins, and the caudal fin may be inclined rather than straight. From this we can deduce an overall evolutionary trend for later lungfishes to more closely resemble the younger ancestral forms, resulting in the final lineage actually looking much like larger versions of Devonian lungfish larvae (this process is called paedomorphosis). Similar changes in the cheek and skull roof bones of lungfish reinforce the belief that paedomorphosis is a major driving force in the evolution of the group.

The first two-thirds of Earth history is called the Precambrian (meaning "before the Cambrian Period"). During this time, life arose, the atmosphere formed, and the Earth's crust underwent many changes as the surface temperatures cooled and the oceans were formed. The oldest life on Earth is represented by sedimentary mound structures built by cyanobacteria, called stromatolites, as well as the fossilised bacteria themselves. Stromatolite fossils in Africa and Australia have been dated at 3500 million years old, and today living stromatolites can be seen in places such as Shark Bay, Western Australia. Stromatolites and other single-celled life forms dominated the Precambrian.

By about 2100 million years ago, the first fossil cells having a nucleus present (eucaryotes) appeared. This great step for cell-kind was to herald the major genetic changes needed for 'speeding up' the process of evolution.

The next advance, the evolution of multicelled organisms (metazoans) is poorly documented by fossils, although a fauna of multicelled animals including jellyfish, worms, sea-pens and bizarre enigmatic forms, has been found in the Flinders Ranges of South Australia at a place called Ediacara. These animals are dated at about 600 million years old, and from the complexity of the animals it is likely that such creatures may have appeared well before this date. The Ediacara sites are unique in their great

▶ Some of the earliest fossil metazoans (multi-celled animals), have been found in Late Precambrian sandstones, about 600 million years old, near Ediacara, in the Flinders Ranges, South Australia. This is *Dickinsonia*, a flat worm.

diversity of very ancient, yet extraordinarily well-preserved soft-bodied animals, which usually do not fossilise well.

Life really took off in a big way at the dawn of the Phanerozoic Eon, about 540 million years ago. The Phanerozoic Eon contains three major Eras: the Palaeozoic (meaning "ancient life"), the Mesozoic ("middle life") and the Cenozoic ("new life"). Many life forms of the start of the Palaeozoic were preserved as fossils, and these reveal an explosion of diversity in

THE FACT OF EVOLUTION

Evolution has occurred and is occurring all around us. Evolution is the change in genetic make-up inherited by the descendants of maybe only one individual, or a breeding pair, which over successive generations renders the lineage so different from the original species that interbreeding with members of the original species population will no longer produce viable offspring. These genetic changes result in changes of shape and size of structures within organisms that enable them to survive in different environments. The modern study of evolution is based on countless millions of pieces of evidence from the fossil record. It can also be tested by modern genetics, tracing mutations within races or tribes of living creatures, especially insects, although the fossil record is still the best evidence we have for the overall placement in time of the origins and changes within different lineages, allowing us to understand how most modern groups of animals and plants reached their current levels of organisation.

When Charles Darwin published his radical book *On the Origin of Species* in 1859, it heralded the "theory of evolution". Darwin's theory of "natural selection" was able to show how an organism's environment is the driving force for change. However, one thing bothered Darwin about it: the lack of hard fossil evidence. Today, with the success of radiometric dating techniques, and probably a millionfold more fossil specimens in museums and private collections around the world, there are few real gaps in the fossil record. Evolution is now an accepted fact that actually underwrites the scientific credibility of all biological, medical and genetical research going on around the world today. The so-called "missing links" have nearly all been found at the major high levels of animal evolution. The task ahead for modern palaeobiologists is to find and elucidate the relationships of animals and plants at finer levels, and there are still many problems of this nature to be solved. These problems are merely ones of fine-tuning and in no way challenge the overall scheme of the established evolutionary hierarchy.

Evolution can be observed in modern times. The apple maggot fly *Rhagoletis pomonella* was originally native to hawthorn trees of North America but has evolved in the last 200 years with the spread of modern agriculture to adapt to different fruit-maturation periods. Under laboratory conditions it has been shown that an ancestral hawthorn fly takes 68–75 days to mature and breed. Recent races of the fly that infest apple trees may take only 45–49 days to mature and breed, while races that infest dogwood trees may take 85–93 days. All of these times correspond to the exact period it takes for fruit on the host plant to mature. Thus, within 200 years different races of *Rhagoletis* have evolved to adapt to seasonal differences in the timing of fruit ripening.

A cladogram represents the sequence of character acquisition in an evolutionary lineage. Here we see the branching sequence linking the lungfish with the dinosaur as they share many advanced features not seen in the extinct jawless fish — for example jaws, teeth, three or more chambers in the heart, lungs, well-developed paired limbs, and many other anatomical features. Cladograms each inherently imply a new classification based on the branching nodes as higher taxonomic groups.

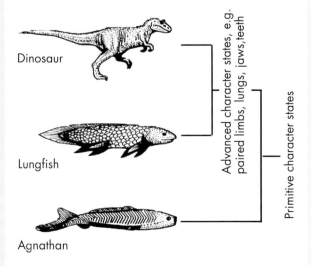

Dinosaur

Lungfish

Agnathan

Advanced character states, e.g. paired limbs, lungs, jaws, teeth

Primitive character states

animals. The fact that many of them had an outer hard protective shell is the reason why the organisms were preserved as fossils and explains why the "explosion of the diversity of life" at this point in time may be explained in terms of a major biochemical change in many types of invertebrates. Most of the modern groups of animals have been around since the beginning of the Phanerozoic Eon, in the Cambrian Period; thus the evolution of many invertebrates, including arthropods (joint-legged animals such as

Another fascinating aspect of modern evolutionary study is the discovery of how developmental changes within the growth of a single individual can explain evolutionary changes. For example as we humans grow to maturity we exhibit many radical changes in the proportions of our head, trunk and limbs as well as late onset changes like increased hairyness. Such changes explain why a human and chimpanzee may have 99 per cent similarity in genes, but differ greatly in appearance. A change in development timing that results in a new species is called *heterochrony*. Different types of heterochronic mechanisms have been identified in various organisms: the retention of juvenile features into maturity may be brought about by earlier sexual maturity (and breeding) or by reducing the rate of growth, and is called *paedomorphosis*; and the addition of extra growth phases to bring about changes in bodily form and structure is termed *peramorphosis*. In most cases the nature of heterochronic change from one species to another may involve several mechanisms, which switch on and off at different stages in the organism's development. Such cases are termed *dissociated heterochrony* and may involve, for example, retention of juvenile characters in the skull (paedomorphosis) whilst the limbs are growing faster and adding on extra growth stages (peramorphosis). The concept of heterochrony is really the new key to understanding evolution—by observing the vast degree of morphological change within the growth of a single species, we can now link such changes to the large-scale evolutionary radiations of many species that came about over millions of years.

Modern molecular biology has breathed new life into old bones. It has added a new dimension to interpreting the evolutionary lineages of animals and plants. By measuring the degrees of similarity between animal proteins or DNA in different species, scientists can estimate a relative time of divergence; and in many cases these match the times suggested by the fossil record and pinpointed by radiometric dating methods. Thus the true test of fossils has been confirmed. A recent example of this is Martin, Naylor and Palumbi's study of the evolution of sharks (1992, published in the journal *Nature*), which showed that similarities in mitochondrial DNA in living sharks suggests divergence times for shark families that tied in perfectly with the known fossil appearances of the groups.

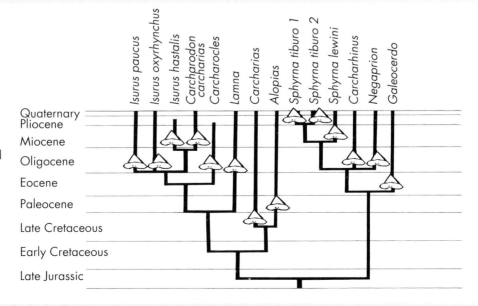

Mitochondrial DNA divergence times of living shark lineages can be tested against the first appearance of fossil forms (here denoted by the shark's tooth icon). In this case there is a very close correspondence suggesting that the fossil record of these sharks gives an accurate representation of their evolutionary history. (After Martin, Naylor and Palumbi, 1992)

DATING ROCKS AND FOSSILS

The Earth is approximately 4.6 billion years old. The evidence for this comes largely from the study of meteorites, rocky remnants from an explosion that occurred in space at the time the solar system was being born. The minerals in these meteorites crystallised from a hot liquid magma during the cooling phase; and once their crystal shape had formed, some of the minerals containing radioactive elements immediately began to decay. By measuring the amount of decay, and knowing the decay rate of the radioactive elements from experiments, we can accurately calculate the time of the formation of the rock. Several different radioactive dating techniques are used, depending on the range of ages being dated. For older rocks up to 4 billion years old examples commonly used are the decay of potassium into argon, rubidium into strontium or uranium into lead. For dating organic materials less than about 50,000 years old we use the natural decay of carbon isotopes.

The commonest method of determining the age of a fossil involves comparing its position in the rock horizon with nearby volcanic rock horizons that have been radiometrically dated. Fossil fish remains near the town of Taggerty in the Cerberean Cauldron complex of central Victoria, in southeastern Australia, have igneous rocks both above and below them, which have been dated using radiometric isotopes, thus constraining the minimum and maximum ages of the fossils. With numerous radiometrically dated sequences around the world, it is now possible to constrain most of the world's outcropping rocks in a similar fashion. Zonations of fossils with known age ranges, based on radiometric age constraints, can therefore be used to tie in the age of other fossils that occur within or near them in layered rock sequences. For example the fossil placoderm *Bothriolepis* is known from all over the world in rocks of Middle–Late Devonian age, although certain species are restricted to much narrower time ranges. This method of assessing the age of a fossil is termed "relative age dating" as opposed to using "absolute age dating" based on radiometric methods. Age ranges of fossil assemblages are termed "biozones". The use of fossils for dating of rocks by comparison with the zonations of known fossil assemblages is called "biostratigraphy". The specific fossils which have characteristic narrow time ranges and are useful for age assessments are called *key taxa* in biostratigraphic schemes.

The radiometric dating of fossils involves measuring the decay rate of a radioactive isotope since it first crystallised from a molten rock. By knowing its constant rate of decay we can then estimate its age by the amount of decay shown within the mineral structure. The clocks represent the half-lives of the radioactive isotope or, the standard length of time by which we can measure how old the mineral is, and thus the age of the rocks in which the fossils were formed.

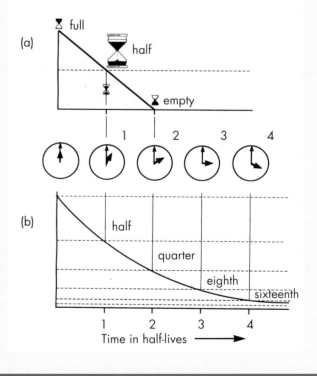

insects, spiders and crabs), echinoderms (spiny-skinned animals such as starfish), cnidarians (corals, jellyfish and sea anemones) and molluscs (clams, snails, squids), took place near the start of the Cambrian. Even the first fish-like creatures, well on the way to becoming vertebrates, had appeared by the Middle Cambrian (see Chapter 1 for the full story).

During the early part of the Palaeozoic Era the first true fishes—those having bone—had appeared and their radiation was in full swing by the late

CLASSIFICATION — AN ORDER WITHIN NATURE

KINGDOM ANIMALIA
(animals, not plants, algae or fungi)

PHYLUM CHORDATA
(chordates, back-boned animals)

SUPERCLASS GNATHOSTOMATA
(jawed vertebrates)

CLASS CHONDRICHTHYES
(sharks, rays and holocephalans)

ORDER LAMNIFORMES
(lamnid sharks)

FAMILY LAMNIDAE
(mackeral sharks, porbeagles, white shark)

GENUS *Carcharodon*

SPECIES *Carcharodon carcharias*
(great white shark)

Linnean classification is based on a hierarchical system of levels, each united by shared anatomical features reflecting a common evolutionary origin.

The system of classifying animals and plants introduced by Swedish naturalist Carl von Linné (also known as Carolus Linnaeus) in the eighteenth century is called Linnean classification. It uses a two-word name system for every species. The first is its generic name (genus), which applies to the group containing a number of different species. The second part is the species name. It enables scientists to identify a species of plant or animal precisely without the confusion of differing popular or common names. Thus the great white shark, or white pointer or white death or whatever-you-want-to-call-it shark, has a binomial name of *Carcharodon carcharias* (genus *Carcharodon*, species *carcharias*). The system extends to other hierarchical levels: families (which contain a number of genera), orders (which contain a number of families), classes (which contain a number of orders), and phyla (which contain a number of classes). There are also many subdivisions between each of these formal levels (superfamilies, subfamilies, infraorders etc.). The rationale behind this hierarchy is that similar organisms can be grouped together, and this grouping is thought to be a natural reflection of their evolutionary relationships.

The human species, *Homo sapiens,* is placed in the Family Hominidae, along with fossil forms such as *Australopithecus afarensis*. The Family Hominidae is placed in the Order Primates, which includes other families such as the Pongidae (containing man's closest living relatives, the chimpanzee and gorilla). Primates and other orders of mammals are placed in the Class Mammalia, which is placed in the Phylum Chordata (including all chordates, creatures having a notochord). This might seem relatively simple, but at the base of the classification are many different levels.

New systems of classifying organisms according to their evolutionary position (cladistic classifications) tend to complicate the matter further. Each time a cladogram expressing one scientist's view of the relationships of a set of organisms is published, then an inherent classification is implied from that view. The classifications used in this book follow conventional usage and do not contain any personal viewpoints.

Silurian, about 420 million years ago. The Devonian Period is often called "the age of fishes" because it was then that they achieved a peak of diversity, and all modern groups of fishes—at least at higher taxonomic levels—had evolved. By the close of the Devonian the first amphibians had appeared, and shortly afterwards, in the Early Carboniferous, the first "protoreptiles".

The great diversification of reptiles into groups such as mammals, dinosaurs and birds took place within the Mesozoic. Dinosaurs and mammals

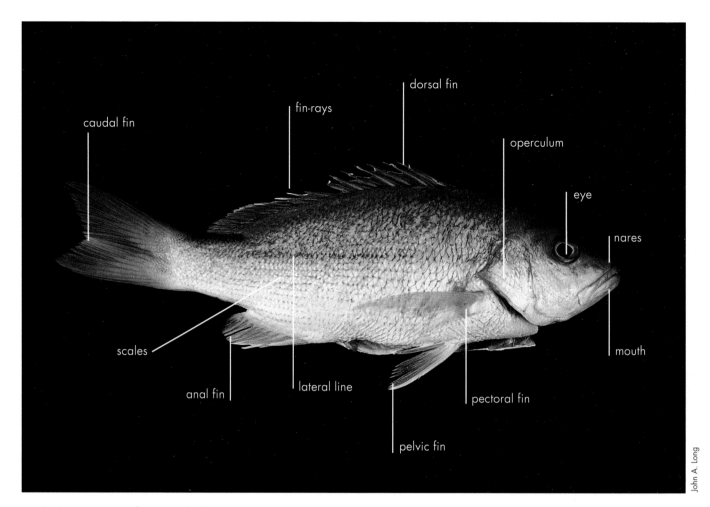

caudal fin

fin-rays

dorsal fin

operculum

eye

nares

scales

anal fin

lateral line

pelvic fin

pectoral fin

mouth

John A. Long

▲ The basic external features of a fish.

Basic skeletal features of a fish (*Eusthenopteron*, a Devonian crossopterygian).

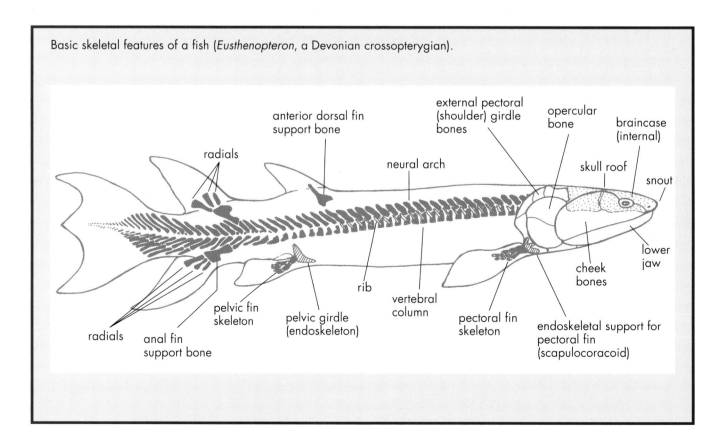

radials

anterior dorsal fin
support bone

external pectoral
(shoulder) girdle
bones

opercular
bone

braincase
(internal)

neural arch

skull roof

snout

rib

vertebral
column

lower
jaw

cheek
bones

radials

pelvic fin
skeleton

pelvic girdle
(endoskeleton)

pectoral fin
skeleton

endoskeletal support for
pectoral fin
(scapulocoracoid)

anal fin
support bone

appeared about the same time in the Middle Triassic, about 230 million years ago, although the latter were suppressed by the reign of dinosaurs for the next 150 million years. The first bird that we know of, *Archaeopteryx*, appeared about 155 million years ago, and by the close of the Mesozoic at least 60 different species of birds had evolved. Dinosaurs ruled the land during the Mesozoic and giant aquatic reptiles such as the plesiosaurs, ichthyosaurs and mosasaurs flourished in the oceans. At the close of the Mesozoic, 65 million years ago, the dinosaurs became extinct along with many other groups of animals and some plants.

In the Cenozoic Era mammals radiated into the niches left vacant by the demise of dinosaurs and large marine reptiles, while the surviving reptiles and amphibians stayed more or less as they were through to recent times. Birds, however, underwent much evolution and radiation during the Cenozoic. The origin of humans (genus *Homo*) lies in this last phase of evolution, somewhere in Africa about 5 million years ago, but that is a story told elsewhere in many books. Our story, the rise of fishes, is one of major radical change in early vertebrate structure, not merely fine-tuning of existing anatomical patterns.

A GUIDE TO BASIC VERTEBRATE ANATOMY

This section will serve as a basic guide to the anatomy of fish fossils discussed in the rest of the book. The vertebrates are a rather conservative group of animals in their basic anatomical structure. Whether you are a bird, a fish or a human, you share the same basic pattern of bones in the skeleton, and the same basic soft tissues, although some modifications are unique to each group of vertebrate. The basic plan of a fish—in having a head with internal skull bones, and pectoral and pelvic girdles—can be seen in an amphibian or a mammal, although the fusion or loss of some bones may characterise many of the higher groups of vertebrates. Vertebrates having four limbs with digits on the hands and feet are generally termed *tetrapods* (amphibians, birds, reptiles, mammals), and a subgroup of these that lay a hard-shelled egg or have their embryos fed from an amniotic sac are termed *amniotes* (reptiles, birds and mammals).

The skeletal system supports the animal and protects its vital organs. It can be divided into a cranial skeleton (the head bones) and a postcranial skeleton (backbone and limb bones). Within the skeletal systems of fishes there are bones formed in the dermis of the skin (dermal bones), bones formed internally

from cartilage framework (endochondral bones) and those formed from precipitation of thin bony laminae around soft tissues embedded in cartilage (perichondral bones).

In fishes the skull is a very complex bone. In some agnathans it is composed of a single ossified unit, but in advanced fishes and in tetrapods it generally has several complex functional units. The braincase (neurocranium) forms around the brain and soft tissues, protecting it and acting as an anchorage site for gill arch bones and the muscles of the head and trunk. The outer layer of dermal bones that cover the head comprises the skull roof bones, the cheek bones, the bones forming the jaws and the bones covering the gill chamber (operculogular bones). The jaws primitively ossify from two cartilages (Meckel's cartilage in the lower jaw, palatoquadrate in the upper jaw) and in osteichthyans and tetrapods are then invested with numerous patterns of dermal bones; primitive jawed fishes may only have the Meckel's cartilage or palatoquadrate ossified.

The bodies of fishes are supported by a stiff fibrous notochord, which is often replaced by a vertebral column made up of a series of backbones (vertebrae) in most fish and tetrapod groups. The axial skeleton includes the vertebral column and fin support bones. In addition there are girdles or rings of bone that support the limbs and act as attachment sites for muscles and ligaments. The front limbs of a fish are called the pectoral fins, and the ring of bones supporting each of those fins is called the pectoral girdle. There are internal bones of the girdle (endogirdle; called the scapulocoracoid in the pectoral girdle) and external dermal bones (exoskeletal girdle). In tetrapods such as humans the pectoral fins have evolved into arms, and the pelvic limbs or fins are now legs. The pelvic girdle is often simple in fishes, but becomes quite complex in tetrapods. Fishes have a well-developed tail or caudal fin, which provides the thrust to propel them through the water. The tail may take on many shapes and forms and may even characterise certain groups of fishes (for example, most coelacanths have tufted tails), but despite this it is really only in the advanced ray-finned fishes, the teleosteans, that complex internal fin skeletons evolved (see Chapter 7).

Aside from the bony skeleton, most fishes have a covering of bony scales. Like the dermal bones of the head and shoulder girdle, the scales form in the dermis and are part of the dermal skeleton. Scales are seen in

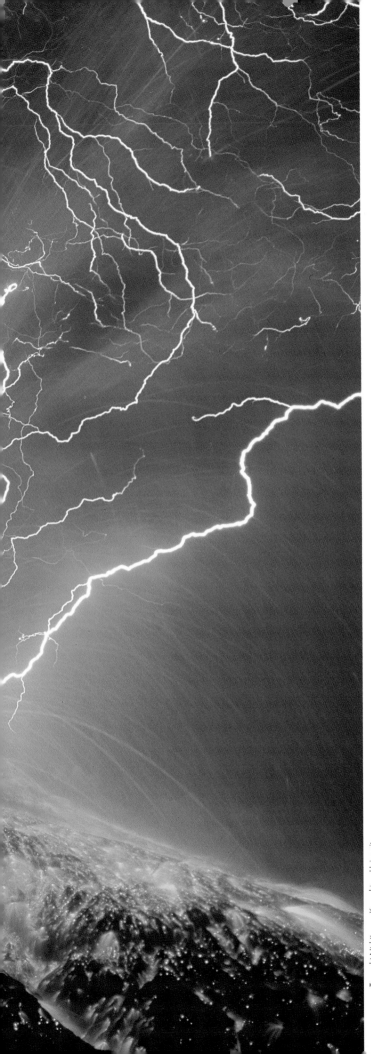

Tsuyoshi Nishiinoue, Kagoshima University

many shapes and sizes and are characteristic for each fish group. Many primitive jawless fishes have simple platelets of ornamented bone forming the scales, and these do not overlap or imbricate. Primitive jawed fishes, such as sharks, have many thousands of small (microscopic) tooth-like scales set in the skin. This gives a sandpaper-like feel to the skin. Each of these placoid scales has a bony base and a crown made of tooth-like tissues (dentine, enameloid). In the osteichthyan fishes there are larger scales that overlap with each other and have fine ornament on the externally exposed regions. The scales give support and protection to the body and carry sensory organs such as the lateral line canal.

The sensory systems of fishes include the eyes (vision), nostrils (olfaction or smell) and the lateral-line system (sensing motion, chemical changes or electric fields within the water). The lateral-line system includes the main sensory-line canals or grooves set in the skull bones and in the scales of the body, as well as pit-line organs which appear as more deeply incised grooves on certain skull, cheek and jaw bones. The lateral lines of fishes give them the amazing coordination needed when a dense school of fish suddenly all turn together and avoid crashing into one another. On the fish itself the lines consist of canals in which lie many small sensory organs that detect subtle changes in water direction or small amounts of chemicals. Special sense organs are found in some groups—such as the ampullae of Lorenzini found in the large snout pores of sharks and rays. These are finely tuned to detect weak electric fields given out by prey items. They enable some sharks to catch fish that have buried themselves just below the sandy bottom of the seabed. Unusually shaped sharks, such as the hammerhead, may have evolved their wide faces to increase the area of ampullae in the head, thereby giving them a more powerful sensory-system.

When we study the fossils of fishes we are generally looking at skeletal anatomy, although the complex bony shields of early fishes often preserved delicate structures of the braincase or jaws, enabling scientists to make accurate reconstructions of some of the soft tissues. However, in most cases, where this kind of three-dimensional preservation is lacking, we generally study the external form of the fish and the shape and structure of what skull bones might be

◀ Volcanic eruptions demonstrate the dynamic nature of the Earth's plates in motion.

visible. Despite this we have a remarkably clear and detailed picture of fish evolution, and new techniques of studying old fossils are yielding more and more information about crucial steps in the evolutionary rise of fishes, from simple jawless wonders to the ancestors of the first land animals.

The Earth's crust is a dynamic system, continually in motion, even though the degrees of movement are impossible to detect with the human eye. Plates of the Earth's crust that carry the continents are continually moving, thrusting their edges underneath or above other plate margins as new crust is being continuously formed by volcanic processes opening onto the ocean floors or erupting along the active margins of continents.

continental crust

volcano

sediment wedges

magma chamber

oceanic crust

spreading centre

subduction zone

convection currents

upper mantle

melting of oceanic crust

lower mantle

oceanic crust

spreading centre

continent

continental crust

volcano

lower mantle

melting of crust

currents in the mantle

upper mantle

DRIFTING CONTINENTS AND FISH EVOLUTION

One of the most important concepts to grasp when trying to understand the complex evolution of species over hundreds of million of years is that the Earth has been continually changing throughout this time. The solid surface layer of the Earth, the crust, is about 5 km thick over the oceans (called oceanic crust) and up to 150 km thick over the continents (continental crust), and it floats on the mantle, a zone of molten rock. The continents and oceans rest on plates, which are in motion. The leading edges of some plates are slowly pushed under or subducted below the overriding margin of neighbouring plates (in subduction zones); and as they are pushed deep into the mantle, they melt. In the middle of the major oceans new crust is being formed at spreading centres, where volcanic rocks well up from the mantle and are added to the margins of plates, which then spread away from the volcanic rift zone.

Volcanos, the surface expression of molten rock reaching the surface of the Earth, form near plate margins where there is active subduction, and the melting of the underthrust plate provides new crustal material that rises up through fractures in the crust to feed the volcanic vents. At collision zones, where two plates meet, the edge of the overthrust plate is forced upwards, forming a belt of mountains. Erosion by wind and water gradually weathers away the upthrust regions, and the tiny rock fragments become sediments that accumulate in lowland basins and depressions, both within continents (in rivers, lakes) and on the seabed. It is here that fossils are preserved.

The study of reconstructing past continental positions and their geographical features is called palaeogeography. The continents we know today all had different shapes and positions on the globe, and many were joined to form supercontinents, such as Laurasia, the great northern landmass (containing North America and most of Asia), and Gondwana, the great southern landmass (containing Australia, Africa, South America, India, Antarctica and many other countries at differing times). The movement of these continents and the changing atmosphere, together with constantly changing sea levels and surface temperatures causes a great number of climatic variables that drive the evolutionary mechanisms. If creatures evolve and are well adapted to life in their environment, then the pressure for them to change is very low. However, their paradise is never constant as changing environmental factors eventually lead to pressures on the organism to adapt to the changes or die out. For example, a plant living near the seashore would have to adapt to increasing salinity if the sea level rises slowly— or else become extinct.

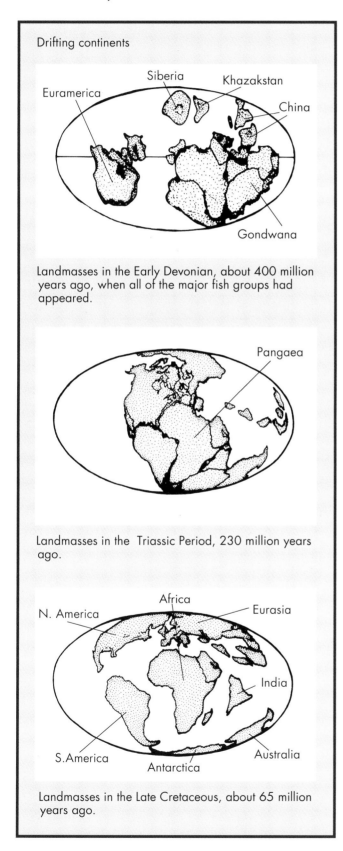

Drifting continents

Landmasses in the Early Devonian, about 400 million years ago, when all of the major fish groups had appeared.

Landmasses in the Triassic Period, 230 million years ago.

Landmasses in the Late Cretaceous, about 65 million years ago.

Over millions of years the movement of continents often resulted in tectonic plate collisions that formed new mountain belts; allowed seas to inundate previously dry lands; joined one landmass with another; and created chains of islands within the seas. Thus species could invade new lands and compete with the existing plants and animals for food resources. Many of the great extinction events in past geological times may have been caused by "armies" of invading species going into a new territory, or major fluctuations in sea level and changes in climate causing a landmass to become suddenly flooded or a lush forested area turned rapidly to desert.

The diagrams shown here give a clear picture of the changing continents. Much of fish evolution was located in waters around Gondwana, the species then migrating via shallow seaways to new regions. When we find, in one region, the oldest and most primitive fossils of a group ("primitive" meaning least-changed from the ancestral stock), it suggests that the group may have originated there. Subsequent findings of more advanced species away from the proposed centre of origin can define a biogeographic pattern and suggest that an historical migratory event may have taken place in past geological times.

A feature of the early evolution of fishes in the Devonian Period is their restriction to certain areas. Groups of organisms that inhabit a restricted space are termed *endemic*, and those which occur all around the globe are termed *cosmopolitan*. Kangaroos are endemic to Australia and New Guinea, whereas seagulls have a cosmopolitan distribution. The discovery of endemic groups of fossil fishes in the Devonian has greatly contributed to refining our picture of the globe and the continents in those times. This information allows geologists to use fish fossils from certain regions to give an estimate of the age of rocks in which they are found, based on knowing the ages and restricted distribution of certain fossil fish species. An example of this is that in China today we have discovered a highly endemic Early Devonian fauna of fishes (more than 95 per cent of those species and genera are endemic to China). Scientists working in northern Vietnam have found similar species in rocks not far from the Chinese border and can now use the data from the Chinese discoveries to deduce that northern Vietnam was once part of the ancient continent of South China and that the new discoveries are Early Devonian in age.

GEOLOGICAL TIME CHART AND SUMMARY OF FISH EVOLUTION

Eon	Era	Period	Epoch	million years ago	Major events in evolution
PHANEROZOIC EON	CENOZOIC	QUATERNARY	RECENT		
			PLEISTOCENE	1.6	Many extinctions
		TERTIARY	PLIOCENE	5	◄ Appearance of hominids
			MIOCENE	25	
			OLIGOCENE	45	Modern mammal families
			EOCENE	57	evolve and diversify
			PALEOCENE	65	
	MESOZOIC	CRETACEOUS		135	Many extinctions; Birds diversify; Main group of mammals have evolved
		JURASSIC		205	◄ First birds; Teleosts diversify — First teleost fishes; Dinosaurs diversify
		TRIASSIC		250	◄ First dinosaurs and mammals
	PALAEOZOIC	PERMIAN		290	Many extinctions
		CARBONIFEROUS		355	First mammal-like reptiles; Amphibians and reptiles diversify; ◄ First amniotes — primitive reptiles
		DEVONIAN		410	◄ Many extinctions; peak of global sea-level rise; ◄ First land vertebrates — amphibians; ◄ Earliest insects (springtails); ◄ Oldest lobe-finned fishes — lungfish and crossopterygians; The age of fishes — fishes diversify into many different groups; ◄ Oldest bony fishes — osteichthyans, actinopterygians
		SILURIAN		438	◄ Oldest tracks on land— Kalbarri eurypterids; oldest vascular plant and terrestrial arthropod communities; ◄ First fishes with jaws — gnathostomes; ◄ Oldest shark scales; Many extinctions
		ORDOVICIAN		510	◄ First fishes (jawless forms) — first bone
		CAMBRIAN		540	◄ Origin of protovertebrates — creatures with a notochord; ◄ Abundant hard–shelled organisms
	PRECAMBRIAN			4500 million years ago	◄ First assemblages of invertebrates at Ediacara; ◄ First cells with a nucleus 2.1 billion years ago; ◄ Oldest life on earth — 3.5 billion year old stromatolites

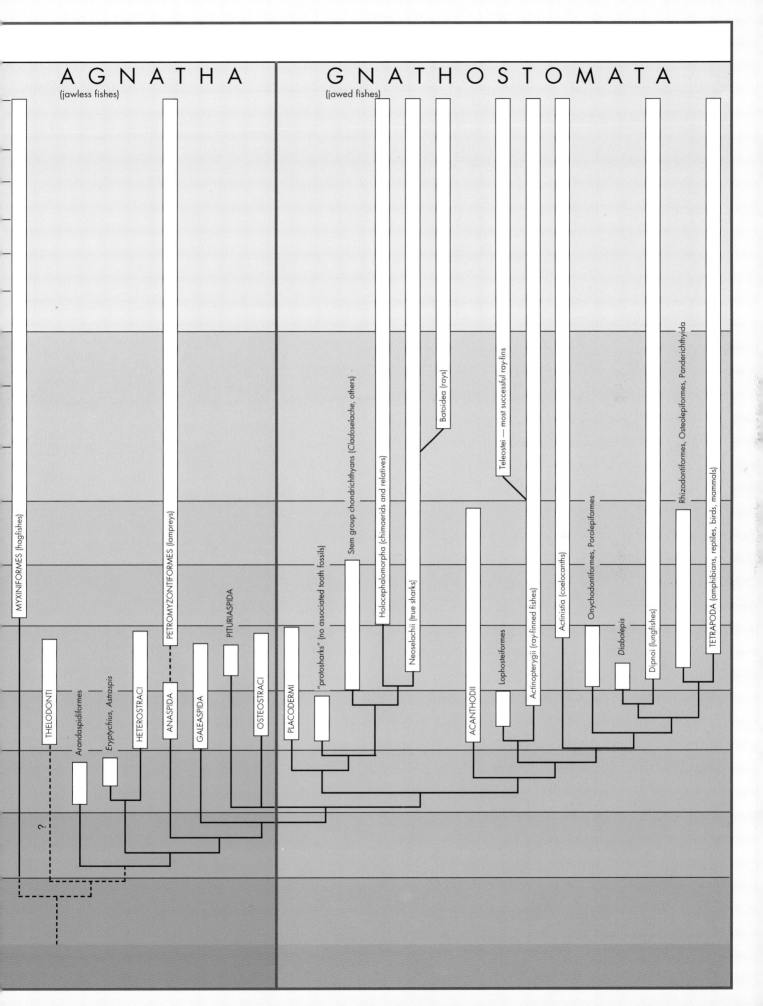

A G N A T H A
(jawless fishes)

G N A T H O S T O M A T A
(jawed fishes)

MYXINIFORMES (hagfishes)

THELODONTI

Arandaspidiformes

Eryptychius, Astraspis

HETEROSTRACI

PETROMYZONTIFORMES (lampreys)

ANASPIDA

GALEASPIDA

PITURIASPIDA

OSTEOSTRACI

PLACODERMI

"protosharks" (no associated tooth fossils)

Stem group chondrichthyans (*Cladoselache*, others)

Holocephalomorpha (chimaerids and relatives)

Neoselachii (true sharks)

Batoidea (rays)

ACANTHODII

Lophosteiformes

Actinopterygii (ray-finned fishes)

Teleostei — most successful ray-fins

Actinistia (coelacanths)

Onychodontiformes, Porolepiformes

Diabolepis

Dipnoi (lungfishes)

Rhizodontiformes, Osteolepiformes, Panderichthyida

TETRAPODA (amphibians, reptiles, birds, mammals)

?

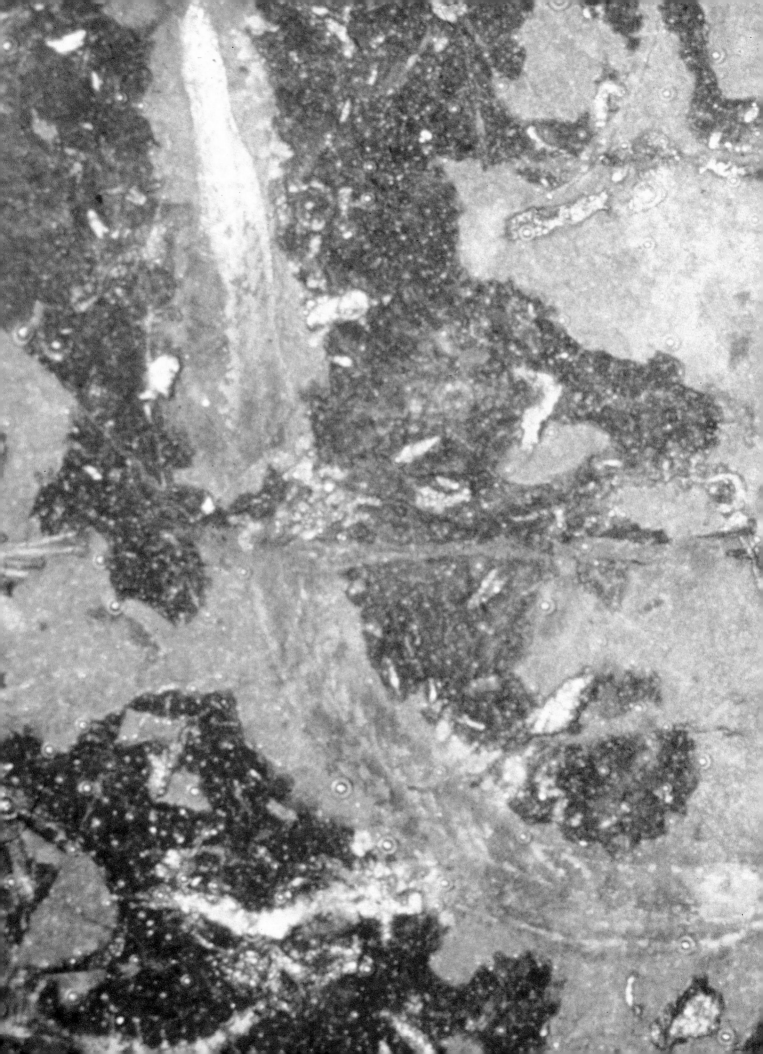

Chapter 1

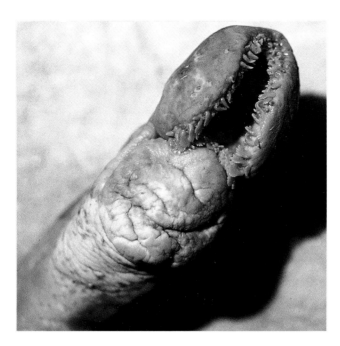

ENIGMATIC FOSSILS: HEADS OR TAILS?

THE ORIGINS OF FISHES

The oldest ancestors of the fishes date back to one small worm-like beast called *Pikaia,* from the Middle Cambrian of British Columbia, Canada. *Pikaia* is the first known creature to have a notochord, comprising a rod of stiffened fibrous tissue supporting the axis of the animal, and also has a tail fin supported by rods of cartilage, another vertebrate characteristic. Other enigmatic creatures having some of the features unique to vertebrates also occur in Cambrian and Ordovician rocks, heralding the eventual arrival of true vertebrates. Yet the transition from spineless invertebrates to the first backboned fishes is still shrouded in mystery, and many theories abound as to how the changes took place. Some argue that echinoderms (the phylum in which modern starfishes are classified) are the possible ancestors of the first fishes, while others champion the sea-squirts (tunicates) and their allies as playing the major role in vertebrate origins. Recent fossil discoveries of the conodont animal, once known only from enigmatic jaw-like fossils, indicates that these creatures were anatomically very close to the first fishes. Whatever the ancestor, the evolutionary steps leading to the first fish are clearly seen in the fossil record, even if the stages belong to many different players.

To most of us, a fish is simply a creature with a bony skeleton that swims, has fins and gills, and is quite often nice to consume with a fine chilled white wine. But all of these characters can be found in creatures that are not fishes; and certain fishes do not have all of them. For example, a Mexican walking "fish" (axolotl) has gills and arms and legs, and is actually an amphibian, not a fish. Sharks and lampreys lack a skeleton of bone (having cartilage instead), and many primitive fossil jawless fishes do not have fins of any kind, only a very simple tail. But all fishes have gills, and all have a notochord, a stiff fibrous rod of tissue that supports the backbone and often disappears after the formation of the bone or cartilage units. The notochord also exists in many primitive fossil ancestors of the first fish, and creatures having this feature are termed chordates and are placed in the Phylum Chordata. Chordates thus include all backboned animals (vertebrates) as well as several primitive creatures that share certain "advanced" anatomical features with fishes.

One characteristic of chordates is the presence of a series of V-shaped bands of muscles along the body, dividing the tail into segments (although the tail in humans is secondarily lost in the evolution of apes from monkeys). These segments are called somites, and the close corresponding numerical relationship between vertebrae and somitic muscles is a feature of all chordates. Another chordate characteristic is the presence of mesoderm, a tissue formed in the embryo that develops into bone and muscle types found only in vertebrates. Perhaps the most significant characteristic of most chordates is the ability to secrete phosphatic hard tissues—including that most advanced tissue of all, bone. Aside from the fishes and higher vertebrates, the primitive groups of animals containing these characteristics include the urochordates (tunicates, or sea-squirts), cephalochordates (lancelets and related forms), and the conodonts (an extinct group known

◀ *Previous pages:* Conodont animal, *Clydagnathus,* from the Carboniferous of Scotland.
Dr R Aldridge, Leicester University.

Inset: A parasitic lamprey, *Geotria australis.* Today's Agnathans are known only from lampreys and hagfish, rather than from the many bizarre forms of the Early Palaeozoic Era.
John A. Long

▲ Sea squirts, or tunicates, are close relatives of the vertebrates. In the adult phase they have a large pharynx with gill slits.

mainly from their phosphatic microfossils). In addition there is a mixed bag of bizarre early fossils simply termed "problematica", some of which could have affinities to the vertebrates. All of these creatures are found in the sea, and this is undoubtedly where the first great evolutionary steps towards higher vertebrates took place.

TUNICATES (UROCHORDATES)

The tunicates are small marine creatures which are commonly called "sea-squirts" because when you pick up these blob-like creatures they often squirt water out. This ability to squirt is because they have well-developed muscles. They feed by taking water in through the mouth and filter it for food items by passing the water through their gill slits. The name tunicate comes from the fact that they are embedded in a tough outer tunic of cellulose (the substance that gives plants their internal support). Despite the lack of similarity between an adult sea-squirt and a fish, it is the sea-squirt's juvenile phase, or larva, that closely resembles a primitive fish. The larva of a tunicate is rather like a tadpole in having a long muscular tail. It is supported by a notochord and has a spinal nerve cord. The head end has various sensory devices to enable the creature to swim and keep its bearing with respect to gravity and the direction of light. On finding a suitable place to settle, it anchors itself by means of three hair-like sticky structures on the head, called papillae, and begins its metamorphosis into the adult form. The adult remains immobile, and as it develops it resorbs its long tail for nourishment. Tunicates are hermaphroditic; that is, they can develop into either sex as they mature and may even reproduce asexually by budding off new animals.

Fossils thought to be of early tunicates are known from the dawn of the Palaeozoic Era, although there is still much debate among scientists as to whether they really are tunicates or belong to completely new groups. One such example is called *Palaeobotryllus* from the Upper Cambrian of Nevada. Its bubble-like form makes it closely resemble the modern colonies of the tunicate *Botryllus*. Also microscopic platelets of enigmatic creatures have been compared with the spicules found in the tunic of modern tunicates and are thus thought to represent ancient sea-squirts.

A problematic form from the Early Ordovician of China, described by Clive Burrett and me, has a phosphatic tubular exoskeleton with large blisters forming tubercles on the inside of the tube. We named this form *Fenhsiangia* (after the town of Fenhsiang in Hupei Province, China), and thought it must be somehow allied to the first vertebrates because vertebrate bone is the only tissue to develop tubercles of this kind. Despite this comparison, the tube-like shape of *Fenhsiangia* gives no clues as to the nature of the organism or its lifestyle. It has tubercles on the inside of the tubes, rather than on the presumed external surfaces, and so is mysterious by comparison with all other known vertebrates. Such forms give us a tantalising glimpse into the complexities of developing primitive vertebrate tissues, but tell us nothing of the internal anatomical features of their owners.

Dr Alex Ritchie, Australian Museum

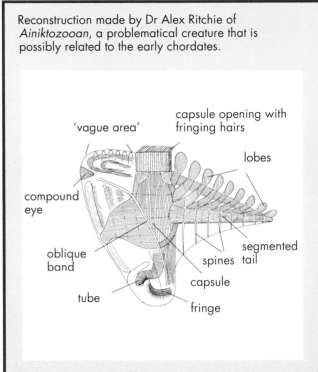

Reconstruction made by Dr Alex Ritchie of *Ainiktozooan*, a problematical creature that is possibly related to the early chordates.

▲ *Ainiktozooan*, a bizarre fossil from the Silurian of Scotland. The creature had well formed compound eyes and a segmented muscular tail. This is just one example of enigmatic fossil creatures that may or may not have played a role in the evolution of the first chordates.

▶ *Fenhsiangia*, an enigmatic fossil from the Ordovician of China. The presence of phosphatic blisters or tubercles suggests that this organism may have belonged to a group ancestral to the earliest vertebrates.

LANCELETS (CEPHALOCHORDATES)

The lancelets have been classically thought of as the closest ancestors of vertebrates because they are small eel-like animals that have well-developed somites with V-shaped muscles, well-developed pharynx with numerous gill-slits, and fin-rays supporting a long median dorsal fin. They are so-named because their bodies are lance-shaped, much like a primitive fish. There are only two known living genera, both of which are marine and reach sizes up to about 7 cm. The adults are sessile and lie buried in the soft sandy shallow sea floor with their mouths protruding above the sediment to take in food from the passing sea water. Lancelets have separate sexes and breed by shedding sperm and eggs into the water. The larvae are very fish-like in having a powerful tail but have fewer gill slits than the adult. The mouth has a circular ring of small tentacles, or cirri, which helps create a current of water around the mouth to enhance feeding.

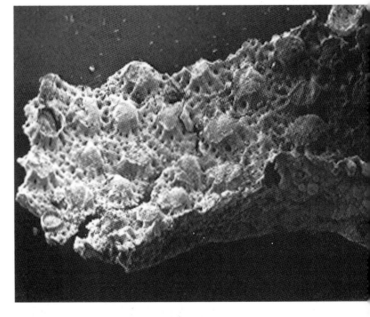

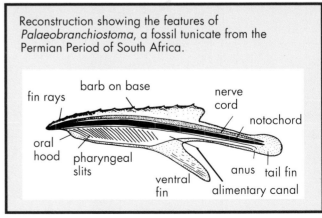

Reconstruction showing the features of *Palaeobranchiostoma*, a fossil tunicate from the Permian Period of South Africa.

Anatomical features of the larva of a sea-squirt.

brain

muscular tail

nerve cord

fin rays

notochord (in green)

gill slits

heart

Although cephalochordates lack a heart they have a blood circulation system that is very close to typical vertebrates in possessing a large central artery, the ventral aorta. The two living genera are *Branchiostoma* (once called *Amphioxus*) and *Epigonichthys*.

The fossil record of lancelets is poor, but none the less includes some well-preserved examples. *Palaeobranchiostoma* is known from the Permian of South Africa, and it closely resembles the living lancelets but has a larger, well-developed ventral or belly fin, and the dorsal fin is also larger and invested with numerous small barbs.

Pikaia, from the Middle Cambrian Burgess Shale of British Columbia, in Canada, could well be the oldest cephalochordate and the most ancient ancestor of all vertebrates. While *Pikaia* has a body form and overall anatomy similar to modern lancelets, it has yet to be fully studied in detail so its evolutionary position remains enigmatic.

Thus the resemblances to vertebrates, as in the tunicates, are seen in the larval stages. By the process of paedomorphosis—retaining juvenile features into a new adult phase, as the juvenile sexually matures earlier—the larvae of either tunicates or cephalochordates could have quite easily developed into a primitive "boneless" fish. The only feature separating them from being a true fish is really the lack of bony tissues.

The next group are more advanced in an evolutionary sense than lancelets or tunicates because they have acquired the primitive phosphatic tissues that come close to being bone.

Douglas Elford, Western Australian Museum

▲ The lancelet, *Branchiostoma*. These primitive vertebrates are filter-feeders which lie buried in the sandy seabed.

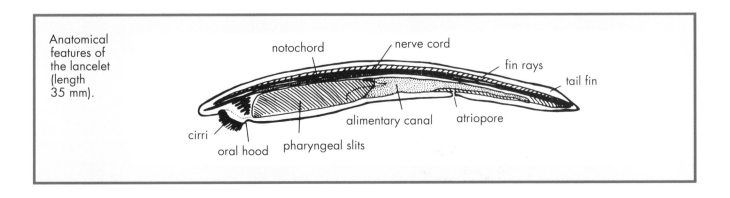

Anatomical features of the lancelet (length 35 mm).

notochord

nerve cord

fin rays

tail fin

cirri

oral hood

pharyngeal slits

alimentary canal

atriopore

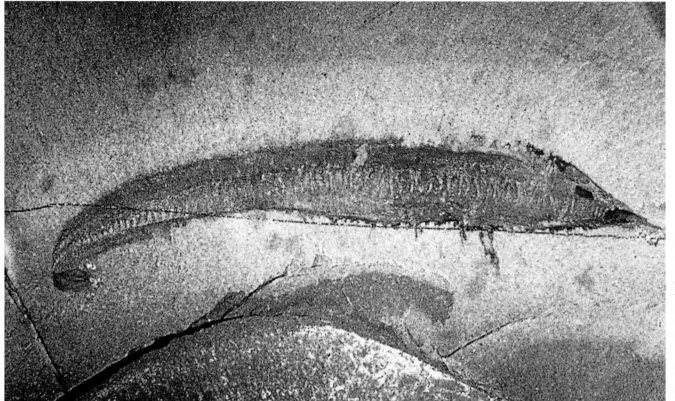

Dr Simon Conway-Morris, Cambridge University

▲ *Pikaia gracilens*, possibly the earliest known chordate fossil, from the Middle Cambrian Burgess Shale of British Columbia, Canada; with a reconstruction below.

▶ The Burgess Shale of British Columbia, Canada, has yielded one of the world's most extraordinarily preserved assemblages of soft-bodied animals dating back to the Middle Cambrian, some 530 million years ago. Besides the many bizarre invertebrates, these rocks have also given us the remains of the earliest possible chordate, *Pikaia*.

▼ Diagrammatic representation of how the first fishes may have evolved from protochordate ancestors.

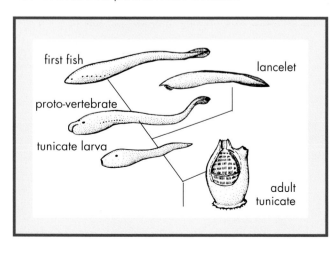

THE CONODONTS: ENIGMATIC FOSSILS

The conodonts (from the Greek, meaning "cone teeth") are an enigmatic group of fossils known principally from tiny microscopic remains of phosphatic jaw-like structures. Countless thousands of these remains can be harvested from Palaeozoic limestones by dissolving the rock in weak acid and sieving off the tiny remains, mostly bone fragments and conodonts. Conodonts have been intensely studied from their isolated remains for correlating and assessing the age of rock sequences, but until quite recently we had no inkling of what kind of creatures owned these enigmatic remains.

Conodonts take the forms of simple rod or cone-like forms, blades with teeth-like protuberances, or complex platform shapes. Studies have shown that sets of these elements often occur together, and thus each conodont animal possessed a number of differently-shaped conodont elements. Studies under the microscope of their tissue type have hinted at vertebrate affinities, and recently a paper published in the American journal *Science* suggested that true bone cells

John A. Long

▲ Conodont fossils (x60). These enigmatic microscopic fossils are commonly used around the world to date the ages of marine sediments. Despite their widespread use, only recently have whole fossils of the conodont animal been discovered. These elements were situated in the animal's head and may have supported filtration or food-sifting structures.

Dr Simon Conway-Morris, Cambridge University

Dr Dick Aldridge, Leicester University

▲ Close up of the conodont animal *Clydagnathus* showing the jaw-like conodont elements in position in the animal's head region. See sketch below.

Sketch showing position of the conodont elements within the head of *Clydagnathus*.

sceral cartilages of eyes

body

conodont elements

1 mm

BONE: ITS DIVERSITY IN FISHES

Nearly all fossil fishes are known principally from the fossilised remains of their skeletons. However, bones as we know them in higher vertebrates (mammals, for example) are quite different from the earliest vertebrate tissues that can be termed "bone". As a general definition, bone is the mineralised tissue supporting the internal or external skeletons of a vertebrate animal. The composition of bone as a living tissue includes the mineralised component, made of phosphatic minerals such as hydroxylapatite, fibres of collagen supporting the formation of mineralised bone, and a degree of vascular tissues which supply blood to and drain blood from the living cells. When speaking of fossil bones, we generally mean just the mineralised component of bone—that

left after the collagen and soft tissues have decayed.

There are two principal types of bone found in early vertebrates: acellular bone and cellular bone. The matrix forming the substance of these bones is essentially the same composition, except that in cellular bone there are cell spaces for osteocytes (bone-forming cells) throughout the bone. Early forms of acellular bone tend to be laminated, or built up in layers, and the bone was laid down from the dermis. Some early jawless fishes have acellular bone (termed "aspidin" in heterostracans), although acellular bone tissues can also be found in the composition of layered bone tissues in more advanced fishes. An example of this is the basal laminated layer of bone in placoderm plates.

John A. Long

were present in some conodonts, although not all palaeontologists agree with this conclusion. Their jaw-like appearance is deceptive for the tooth-like cusps along the ridges never appear to show any sign of wear, as teeth normally do. This means that the structures may have merely been supporting gills or filter-feeding

devices, and not used for food reduction in a direct way.

In the early 1980s the first fossil remains of whole conodont animals were found in the Granton Shrimp Beds near Edinburgh, of Early Carboniferous age, about 340 million years old. They were long worm-like creatures with tails having supporting fin rays,

John A. Long

The structure of primitive agnathan (heterostracan) dermal bone.

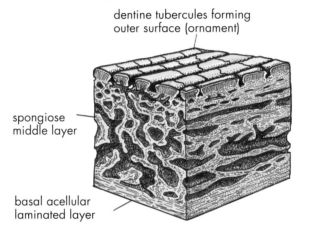

dentine tubercules forming outer surface (ornament)

spongiose middle layer

basal acellular laminated layer

complexity and composition of their external layers, how they grow, and whether or not they have resorption features.

Other types of bone found in early vertebrates include perichondral bone and endochondral bone. Perichondral bone is thinly laminated acellular bone around soft tissues that pass through cartilage. For example, a placoderm that has a cartilage braincase may have perichondral bone layers around the nerves and arteries passing through the wall of the braincase. When fossilised these delicate tubes preserve the outline of much of the soft tissue anatomy, which can be prepared out using weak acetic acid. Perichondral bone also forms in the endogirdles supporting the pectoral and pelvic fins. Endochondral bone is bone formed around a cartilage precursor. In higher vertebrates such as reptiles and mammals this makes up much of the internal skeleton, forming the arm and leg bones. Endochondral bone was not present in primitive fishes, but became a specialised feature of certain groups of jawed fishes (osteichthyans and acanthodians).

In addition to these basic categories of bone, there are many different dental tissues that evolved within the fishes and carried on to higher land animals. Some of these dental tissues form an important part of the skeleton in early fishes. For example in chondrichthyans, which have largely cartilaginous (non-ossified) skeletons, a type of dentine with a thin enameloid layer makes up the dorsal fin-spines. In sharks nearly all of the fossil hard parts are of similar types of dental tissues, including the placoid scales (which actually resemble miniature teeth) covering the body. The dermal bones and scales of many early jawless fishes are similar to this, in having layers of dentine over bone (for example thelodont scales).

Most of the skeleton in early fishes is made of bone formed from the dermis, called dermal bone. This "dermal skeleton" includes all the externally visible bones of the head and trunk, as well as the scales and many of the bones forming the biting surfaces inside the mouth. In crossopterygian and dipnoan fishes the dermal bones and scales are primitively covered by a shiny enameloid layer, which covers an interconnecting network of flask-shaped cavities in the top layer of bone. This complex tissue is called cosmine (see page 131). Other specialised dermal bone tissues developed in certain osteichthyan groups and are characterised by the nature,

and in the head region were found the cluster of little conodont elements. This evidence, together with new data on the "bone" structure of conodonts, is powerful evidence that they were close to the line leading to true vertebrates. If the possession of true bone is sufficient grounds to call a creature a vertebrate, then conodonts could indeed be classified as early fish. However, as they lack many of the refinements we see in the earliest fish fossils—such as dentine layers over sculptured bony plates—I like to place conodonts one peg lower than fish on the evolutionary scale, but this is just my opinion.

THE MITRATES:
STRANGE ALLIES OF THE VERTEBRATES?

Mitrates are a group of echinoderms, the phylum containing starfishes, sea-urchins and sea-slugs. The theory that early fossil mitrates are closely related to the first vertebrates has not gained much favour in recent years, aside from the work of Dick Jeffries of the Natural History Museum, London. However, most workers in the field today regard mitrates as simply an interesting group of echinoderms. They lack bone, a notochord and gills; and resemblances to early vertebrates proposed by Jeffries, such as the muscular tail, are regarded as features evolved within the echinoderms in parallel to vertebrate evolution, rather than as a step towards the lineage leading to true fishes.

BONE: ITS EARLY BEGINNINGS

The identification of true vertebrates in the early fossil record relies heavily on the definition of what exactly bone is. Fishes have bone in which there are cell-spaces for osteocytes (bone-producing cells), and the external layers of the most primitive of all fish bones have an ornament covered by a thin enameloid layer over a dentine layer. Layers of non-cellular bone, found in some fossil jawless fishes, is called "aspidin".

Bone is the key to understanding the success of the fishes. Bone provides a solid support for attachment of muscles, and this gives a greater efficiency for using a muscular tail to propel the creature through water. Faster speeds give escape from predators and the ability to catch slower-moving prey. Bone not only enables the greater locomotory improvements required but also acts as a chemical storehouse for phosphates and other chemicals required in daily metabolism. Furthermore, it gives protection to more vulnerable parts of the anatomy, such as the brain and heart, enabling the organism to have a greater chance of survival after a battle with an attacker. Once bone evolved in its refined state, fishes underwent an explosion of evolutionary diversification.

Despite many claims that vertebrate bone goes back to the Late Cambrian, about 510 million years ago, recent studies have criticised these claims and identified the ancient scraps of "bone" as belonging to the hardened external cuticles of early fossil arthropods. The oldest identifiable fossil vertebrate, one that has undoubted real bone, comes from the Early Ordovician rocks of central Australia, dated at about 485 million years old. These point to Gondwana as the most likely place for the origin of all vertebrates. As we shall see in the next chapter, the oldest well-preserved fossil fishes, known from relatively complete remains, come from Australia and South America, regions in the ancient supercontinent of Gondwana.

The first fishes thus emerge as not much different from the larvae of tunicates, or adult lancelets or the extinct conodont animals. Fish have bony skeletons, and the first fishes are simply "protochordates", like the conodont animals, that have extensive external coverings of bone. This bone is very special in that it has an enameloid surface layer covering a sculptured dentine surface ornament, a spongy layer that housed bone cells and fibrous collagen, and a non-cellular laminated basal layer. The first fishes had only external or dermal bone—bone formed in the dermis of the skin. Internal bony skeletons were to come much later, and this development heralded the next great explosion in fish evolution when jaws and teeth also appeared at the same time.

▲ Reconstruction of a conodont animal, a possible primitive ancestor of the first fishes. After Dr Dick Aldridge.

Fossil mitrate, here shown in top and bottom views. These are actually a bizarre type of echinoderm. Dr Dick Jeffries of the Natural History Museum, London, has argued that these are the closest invertebrate group to the vertebrates.

GILLS

Fishes primitively respired by means of gills, and only advanced bony fishes used lungs as accessory respiratory organs (see the box on the origin of lungs, page 173). The gills are composed of numerous thin-walled soft tissues, called gill filaments, supported by gill arch bones. Gills work as gaseous exchange organs because they present a very large surface area of highly vascularised tissue in contact with the water. The gills take oxygen from the water and release carbon dioxide. The gills may be developed on one side of the support bar and be termed a "hemibranch" or be present as two series of gills on each, making a complete gill or "holobranch". In addition to regular respiration, the gills of some fishes, like lampreys, may supplement kidneys as excretory organs. Some teleosts can even excrete ammonia via glandular cellular elements present in the gills.

Primitively there may be numerous pairs of gills—some agnathans having up to 25 or more pairs (for example, *Legendrelepis*, Late Devonian). In cephalochordates such as *Branchiostoma* the walls of the pharynx have large numbers of vertically oriented gill clefts, each separated by primary bars of a stiff gelatinous substance. These bars have no gill filaments developed, only vascular supply in and out of the bars themselves. In many armoured agnathans, like the osteostracans, the gills are supported from ridges running down the inside of the bony armour. The major distinction between agnathan gills and those of gnathostomes is that in cross-section the agnathan's gills are supported externally, whereas in jawed fishes the gills are attached internally (mesially) to gill arch bones. In lampreys and most agnathans the gills open to the exterior via rounded gill slits, which allow water to pass in and out of the gills. Some fossil agnathans, such as heterostracans (Heterostraci) have a branchial opening which enables flow of water out past the gills. Most gnathostomes have a different system in which water enters by the mouth and is expelled through the gill chamber, exiting through either gill slits or an operculum, a bone covering the gill chamber. In most gnathostomes the gills range from seven pairs in some sharks to typically five in most sharks and bony fishes. The larvae of some fishes have external gills that protrude from outside the gill chamber into the water, but all adult fishes have the gills protected inside the gill chamber.

The gill arch support bones are a complex series in jawed fishes and are generally divided into a ventral series and a dorsal series. In addition to supporting the gill filaments, in osteichthyans they may also support tooth or denticle-bearing bones to aid in food reduction. Some fishes, such as certain acanthodians, bear gill rakers or an additional series of small bones from the primary gill arches, which aid in filtration of food from the water as it leaves the gill chamber. In many lineages of fishes the gills sometimes became modified as primary feeding organs for filter-feeding, such as the whale shark; in this form the gills trap the small food items by gill rakers and the food is then passed down the gill arch back into the gullet.

In chondrichthyans, acanthodians, some placoderms and some primitive osteichthyans the first gill pouch is modified as a spiracle, a small opening behind the eye preceding the full-sized gill slits or gill chamber. In some sharks and rays the spiracle has a valve to regulate the flow of water through the gills or enable the fish to force water over the gills by closing the valve and forcing water back from the mouth. In some cases, water can be spouted out through the spiracle to eject any foreign matter that has entered the mouth.

The evolution of the jaws in fishes is generally accepted as being derived from modification of the first gill arch bones, termed the mandibular arch. The next series of gill arch bones is called the hyoid arch, and it contains specialised bones that are modified in jawed fishes to brace the jaw joint (for example, the hyomandibular and interhyal bones). The remaining gill arches are generally similar in shape but decreasing in size towards the back of the fish. Certain fish groups may have additional specialised gill arch bones, such as "pharyngobranchials" which may point forwards or backwards depending on the group and its functional use.

Chapter 2

JAWLESS WONDERS: THE FIRST FISHES

THE AGNATHA

Jawless fishes (agnathans) have a fossil record spanning nearly 500 million years, the oldest identifiable fragmentary remains coming from central Australia. The earliest recognisable bony shields from these jawless wonders, found in shallow marine sandstones of the Early Ordovician Period, about 450 million years ago, in what is now central Australia. The first completely preserved fishes are about 470 million years old and were found recently in Bolivia in central South America (also part of ancient Gondwana). Shortly after the Ordovician the agnathans radiated into many diverse groups, most of them characterised by bizarre armoured bony shields covering the head, each with their own shape and surface sculpture. The geographic distributions of these highly distinctive groups play a major role in determining the placement of continents on the globe in Palaeozoic times. The early evolution of agnathans involved many of the great advances in vertebrate history, such as the development of cellular bone, paired limbs, intricate sensory-line systems, dentine-like tissues, complex eye muscle patterns, and the inner ear with two semicircular canals. Agnathan fossils are widely used for dating and correlation of Middle Palaeozoic sedimentary rocks, and one group, the thelodonts, are particularly useful in this respect because each individual fish left thousands of characteristic scales in the sediment. In spite of the agnathans' great radiation during the Silurian and Devonian, only the naked-skinned lampreys and hagfishes survived the extinctions, and these constitute today's only living jawless fishes.

Agnathans take their name from the Greek *gnathos* meaning "jaw" and the prefix *a* meaning "without", as they are a group of mostly extinct fishes that lacked true bony jaws and teeth. Today the eel-like lampreys and hagfishes are our only representatives of this once-flourishing group which reached an acme of diversity in the Early Devonian, about 400 million years ago. Although modern agnathans hold little interest to most humans, apart from being a minor food source in some countries, the study of fossil agnathans is vital to our understanding of many important anatomical transformations that took place early in vertebrate evolution. The fossil agnathans are our only window into understanding the origins of jaws, the evolution of cellular bone, and the complete organisation of the standard vertebrate head pattern, and their significance to zoology therefore quickly becomes clear. This aside, it is the simple beauty and mystery of the numerous bizarre-looking agnathan

◄ *Previous pages: Arandaspis, one of the earliest jawless fishes. John A. Long*
Inset: Dr Alex Ritchie at the Arandaspis site in the Northern Territory, Australia. Dr Alex Ritchie, Australian Museum

► Lack of jaws did not inhibit the great diversification of agnathans. Many had elaborate armoured head plates and thick bony tail scales like *Errivaspis waynensis*, an Early Devonian heterostracan, from the Wayne Herbert Quarry, Herefordshire, UK.

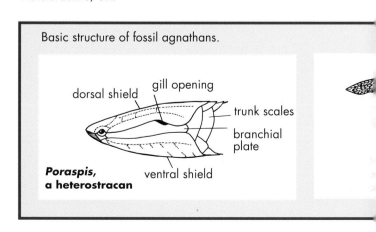

Basic structure of fossil agnathans.

dorsal shield · gill opening · trunk scales · branchial plate · ventral shield

Poraspis, a heterostracan

John A. Long

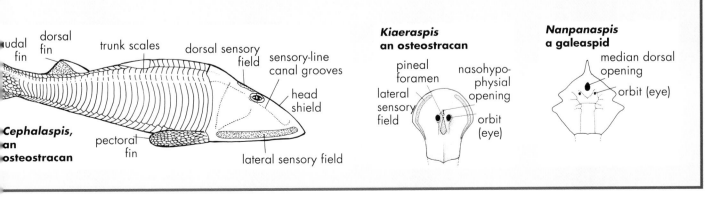

Cephalaspis,
an
osteostracan

udal
fin

dorsal
fin

trunk scales

dorsal sensory
field

sensory-line
canal grooves

head
shield

pectoral
fin

lateral sensory field

Kiaeraspis
an osteostracan

pineal
foramen

lateral
sensory
field

nasohypo-
physial
opening

orbit
(eye)

Nanpanaspis
a galeaspid

median dorsal
opening

orbit (eye)

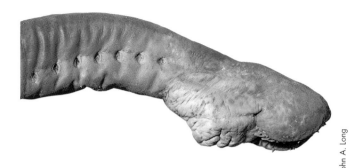

John A. Long

fossils that interests us, as much as their quintessential scientific value.

Agnathan fishes have even played a role in recent human history. King Henry I of England died in Lyons, France, on 1 December 1135, after an excessive banquet in which he gorged himself on lampreys. Lampreys are still considered a great delicacy in some parts of Europe, although only as a minor food source. They are parasitic fishes that feed on other live fishes by attaching themselves with an oral sucker disc. They cut into the flesh to feed on the blood of the host. Their relatives the myxines, or hagfishes, are far more primitive than lampreys in many aspects of anatomy, and are deep-sea carrion feeders. Hagfishes are the largest extant agnathans and may reach lengths of up to 1.4 m.

The modern lamprey and hagfish have fossil records spanning back to the Carboniferous Period, remaining almost unchanged throughout the past 340 million years. The extinct fossil agnathans include six major types of armoured and some non-armoured forms, most of which had evolved by the start of the Silurian Period, about 430 million years ago; these are named the Osteostraci, Heterostraci, Anaspida, Thelodonti, Galeaspida and Pituriaspida. Only three of these groups, the primitive Heterostraci, the Thelodonti and the Pituriaspida, are recorded in Australia and other Gondwanan countries. The Osteostraci are unique to the ancient Old Red Continent Euramerica (Europe, Greenland, western Russia and North America), and fossils of the Galeaspida are found only in the ancient Chinese terranes.

THE OLDEST FISHES: ORDOVICIAN AGNATHANS

The first jawless fishes include the primitive Ordovician forms from Australia (*Arandaspis* and relatives), South America (*Sacabambaspis*) and North America (*Astraspis* and *Eryptychius*). These all share the

◀ One of the living lampreys, *Geotria australis*. Note the numerous rounded gill slits down the side of the head.

▼ Living jawless fishes all lack bony armour and paired fins. The hagfishes are even more primitive in their anatomy than any of the fossil agnathans. Lampreys are now believed to be the direct descendents of the anaspids.

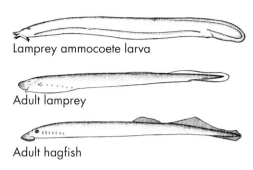

Lamprey ammocoete larva

Adult lamprey

Adult hagfish

primitive feature of having numerous paired openings for the gills, a feature reduced in number in all subsequent jawless fishes except for some anaspids. The bone making up the shields of the North American forms such as *Astraspis* is composed of four layers of phosphatic minerals including fluorapatite and hydroxylapatite. This suggests a close relationship to the heterostracans, a diverse group that had similar shields to these Ordovician forms but which possessed only one branchial opening over the gills.

The earliest relatively-complete fish fossils come from near Alice Springs in central Australia, where they occur in fine-grained sandstones dated at about 470 million years old. When the fossils were first

Reconstruction of the brain (dorsal view) of an advanced fossil agnathan, *Benneviaspis*, an osteostracan. Soft tissue structures can be restored with confidence from the outlines of the nerves and brain cavities, as preserved by thin perichondral bone layers that enveloped the soft tissues within the cartilage parts of the skull. (After the work of Dr Philippe Janvier, Paris)

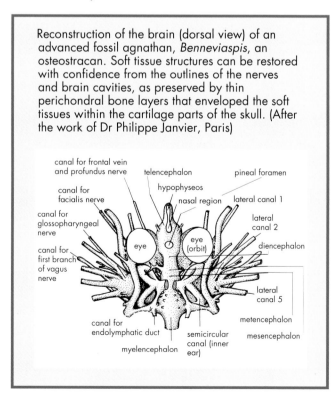

canal for frontal vein and profundus nerve
telencephalon
pineal foramen
canal for facialis nerve
hypophyseos
nasal region
lateral canal 1
canal for glossopharyngeal nerve
lateral canal 2
canal for first branch of vagus nerve
eye
eye (orbit)
diencephalon
lateral canal 5
canal for endolymphatic duct
metencephalon
semicircular canal (inner ear)
mesencephalon
myelencephalon

BASIC STRUCTURE OF PRIMITIVE AGNATHANS

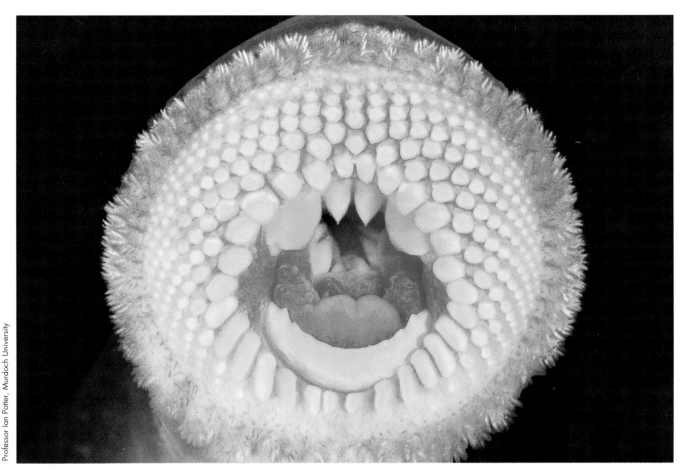

Professor Ian Potter, Murdoch University

▲ Close-up of the oral disk of a lamprey. The lamprey attaches itself to other fishes and rasps away with these teeth, enabling it to feed on the blood of the prey.

Early agnathans are known largely from fossils of their cranial shields and also from isolated scales and bone fragments. In rare cases, whole fossil agnathans are found, showing that the early, heavily armoured forms had tails covered by thick bony scales. Often they have a dorsal fin or an anal fin, or both, but the presence of paired pectoral fins is known in only two groups, the Osteostraci and the Pituriaspida.

The bone forming their shields is made up of an outer ornamental layer, often covered by a shiny enameloid layer over the dentine and perforated by numerous pores. Below this is a middle vascular layer with spaces for collagen fibres and, in some cases, bone cells (osteocytes). The base of the bone is made up of laminated non-cellular bone (sometimes called aspidin). The typical form of the bony shield differs in each of the armoured agnathan groups. The shield in the Osteostraci (meaning "bone shield") and the Galeaspida (meaning "helmet shield") is formed from a single unit of bone perforated by holes for the eyes, nostrils and other sensory organs, and open on the underside for the mouth. In the Heterostraci (meaning "different shield") the armour is formed from several plates of differing

sizes, and the body is covered by large overlapping scales. In the Anaspida (meaning "no shield") and the Thelodonti (meaning "scale tooth") there is no enlarged bony covering, only the scales covering the head and body. These are large, flat scales in some anaspids; but others, such as *Jamoytius* from Scotland, appear to have naked bodies, approaching the condition seen in lampreys.

The braincase is known in some fossil agnathans. This shows that the inner ear possessed two semicircular canals and that the brain was well formed and segmented into discrete divisions (as in higher vertebrates). A large vein drained the blood from the head, and a complex system of sensory fields or lateral lines was developed in most fossil forms. Only in the Galeaspida and the Osteostraci are the soft tissues of the braincase preserved as delicate shells of perichondral (laminar) bone, and these are also the only groups to have the head vein placed dorsally near the top of the armour.

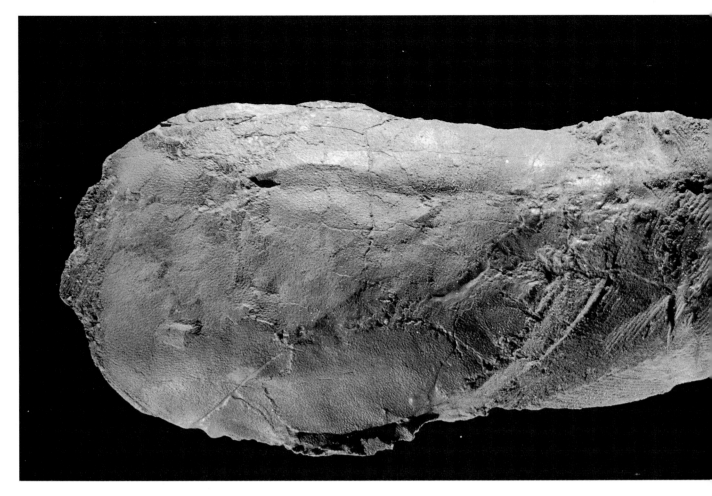

▲ This is the oldest complete fossil fish known. It is *Sacabambaspis*, from Bolivia, of Late Ordovician age. Total length of the specimen is just under 30 cm.

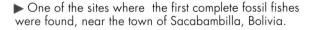

▲ *Arandaspis*, a 470-million-year-old jawless fish found in central Australia. The fossil shown here is the impression of the bony plates preserved in sandstone; the dorsal shield is seen above the ventral shield. The impression of the ribbed clam shell in the front of the specimen approximates to where the mouth would have been in life. Length of specimen: approximately 20 cm.

▶ One of the sites where the first complete fossil fishes were found, near the town of Sacabambilla, Bolivia.

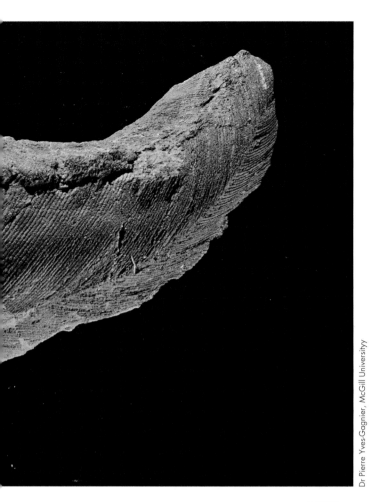

Dr Pierre Yves-Gagnier, McGill University

(belly) shield with up to 14 or more paired branchial plates covering the gills. The eyes were tiny and situated right at the front of the head, like the headlights on a car, and there were two tiny pineal openings on the top of the dorsal shield, probably light-sensory organs. The tail is largely unknown, except for the fact that it bore many rows of long trunk scales, each ornamented by many fine parallel ridges of bone, making them comb-shaped. *Arandaspis* occurs with a number of previously unknown forms of jawless fishes, most of which are still being studied by Dr Ritchie.

The discovery of the world's first complete Ordovician fish fossils in central Bolivia by French-Canadian student Pierre Yves-Gagnier in the mid-1980s caused worldwide scientific interest when preliminary results were published in *National Geographic* magazine. These fish, called *Sacabambaspis* after the town of Sacabambilla in Bolivia, are slightly younger than the Australian fossils (around 450 million years old), but much better preserved. They show the entire articulated armour and body form of the fish. Like *Arandaspis*, *Sacabambaspis* has a large dorsal and ventral shield with numerous rectangular branchial plates, small eyes at the front of the skull and paired pineal openings. It also has a rounded plate on each side near the front of the head. The body is covered with many fine elongated scales, and although it lacks paired or median fins, the tail was quite well developed.

The North American fishes *Astraspis* and *Eryptychius* were long known as the earliest fish fossils, having been first described by Charles Doolittle Walcott in the late nineteenthth century. Their abundant remains come from the Harding Sandstone of Colorado and are preserved as isolated small fragments of bone. Only recently has an almost-complete fish been found. This shows the overall body of *Eryptychius* to look like *Sacabambaspis*, although it has much coarser, rounded

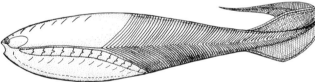

▲ ▼ Reconstructions of the primitive Ordovician fishes *Sacabambaspis* (above) and *Arandaspis* (below).

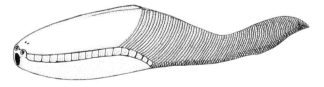

found in the rocks in the mid-1960s the strata were immediately thought to be Devonian age, because at that time Ordovician fish fossils were virtually unheard of. Further collecting at the sites by Dr Alex Ritchie of the Australian Museum in the 1970s and 1980s has yielded a number of good specimens of these early primitive fishes. The shields of *Arandaspis* (named after the Aranda tribe of Aboriginal people) are not preserved as bone but as impressions in the ancient sandstones. These tell us exactly the shape of these armoured agnathans and what their body scales were like. *Arandaspis* has a simple dorsal (top) and ventral

▼ *Astraspis* is one of the oldest vertebrates known from North Amerrica, and was initially described from fragmentary plates of bone from the Late Ordovician Harding Sandstone of Colorado by Charles Doolittle Walcott. In recent years more complete material has been found enabling this new reconstruction of the fish. (After the work of Dr David Elliott)

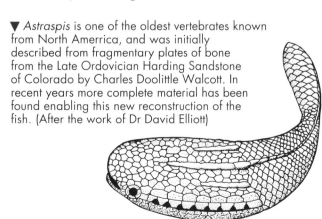

scales covering the tail, and its shield is made up of many polygonal units, called tesserae. The significance of these fossils lies in their excellent bone preservation, and they have played an important role in discovering how primitive bone evolved. The bone of *Astraspis* has four layers: an outer thin layer of enameloid capping a second layer of dentine, which forms the ridges and tubercles of the plates, a third layer of of cancellous or spongy bone, and a fourth basal layer of aspidin, a layered hard tissue that lacks bone cells.

Although these primitive Ordovician agnathans closely resemble the Heterostraci (one of the major Silurian and Devonian groups of agnathans), they lack one distinct feature of that group—a single external branchial opening for the gills. In many other respects, such as having a shield formed of numerous plates, and similar bone structure, they are very close. The heterostracans most likely evolved from such ancestral stock.

▶ Shield of *Anglaspis*, a cyathaspid from the Early Devonian of England.

▼ This is a cornual plate of a large agnathan, *Ganosteus*, from the Late Devonian of the Russian platform. These giant jawless fishes may have reached lengths of nearly 1 m, and were the last surviving family of heterostracans.

THE HETEROSTRACI: A GREAT RADIATION

The Heterostraci (commonly known as heterostracans) are a diverse and wonderful assemblage of ancient jawless fishes. They are easily recognised as fossils by the several plates of bone forming the shield, their single branchial (gill chamber) opening, and their often characteristic elaborate bone surface patterns. Typically there are two large bony plates covering the top and underneath of the head (termed the dorsal and ventral shields), and a single plate covering each of the gill openings

John A. Long

Dr Oleg Lebedev, Palaeontological Institute, Moscow

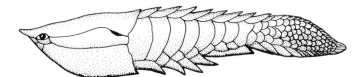

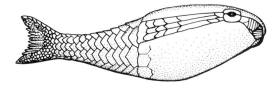

▲▼ Reconstructions of *Anglaspis* (top) and *Athenaegis* (bottom), two primitive heterostracans.

John A. Long

▲ *Lyktaspis nathorsti*, an unusal pteraspid with a long rostrum from the Early Devonian of Spitzbergen (cast).

▼ *Athenaegis*, a complete cyathaspid heterostracan from the Silurian of the Northwest Territories of Canada (length approximately 5 cm).

along the side of the armour (the branchial plates, or plate, where they are fused into one bone). Smaller separate plates may form around the eyes (such as the orbital, suborbital and lateral plates), and the mouth may also have unusual oral plates lining its opening. Sensory lines criss-cross the plates as linear or curved grooves in the bone or as visually clear lineations between the surface ornamental ridges and pustules.

The heterostracans underwent their major radiation early in the Silurian Period, and were common in Euramerica and Siberia throughout the Devonian. The largest heterostracans were the giant flattened psammosteids, reaching estimated lengths of 1 m or more, but most heterostracans were small fishes about 10–15 cm in total length.

The small traquairaspids and cyathaspids from Arctic Canada and Britain are easily distinguished by their relatively simple-shaped shields with highly elaborate surface

ornament. They are primitive heterostracans that lack the elaborate spines developed later and the tail has only a few large scales. Some of these such as the Silurian *Athenaegis* (the name meaning "Athena's shield"), from the Delorme Group of the Northwest Territories of Canada, are beautifully preserved as whole fishes. *Athenaegis* was a small fish about 5 cm long, which had a V-shaped leading edge on the lower lip of the mouth that may have been used for plankton- or detrital-feeding.

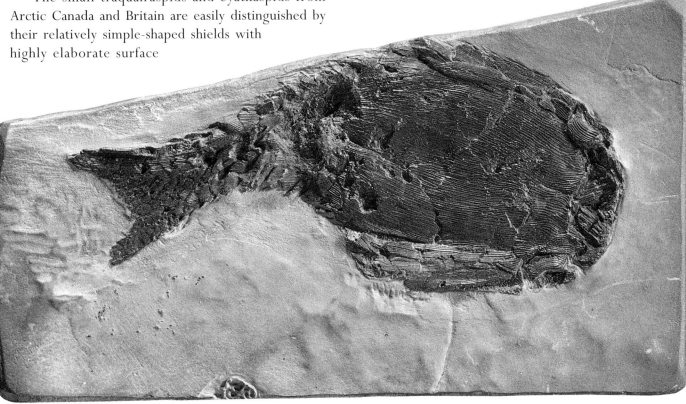

Dr Mark Wilson, University of Alberta

John A. Long

▲ Shields of the pteraspid *Rhinopteraspis* from the Early Devonian of North America.

▶ Reconstruction of *Drepanaspis*, a large flattened heterostracan from the Early Devonian of Germany, seen in side view.

White) from Britain and France, and the large flattened form, *Drepanaspis*, from the black Hunrucksheifer shales of the Rhineland, Germany.

Other groups of heterostracans having strange armours include a group unique to the Russian terranes, the amphiaspids. These had wide, rounded armours made of a single piece of bone, and some of the shields look a bit like flying saucers. Most had shields about 10–18 cm long, the largest forms being as long as about 40 cm. *Lecianaspis* and *Elgonaspis* had bony feeding tubes or scoops at the front of the head that may have functioned as a pump to suck in small organisms from the mud. The eyes were very small or entirely absent in the amphiaspids, as they lived in muddy habitats and instead relied more on their lateral-line

Other cyathaspids, such as *Traquairaspis*, *Corvaspis*, *Tolypelaspis* and *Lepidaspis,* have very distinctive surface ornament, consisting of many polygonal units of elaborately sculptured bony ridges. The cyathaspids flourished during the later half of the Silurian Period but were extinct by the early part of the Devonian.

One of the most successful groups of Devonian heterostracans were the pteraspids (Greek *pteros*, meaning "wing", *aspis* "shield"), so-named because of their wing-like pointed spines at the sides of the armour, called cornua. Pteraspids have a more complex shield than cyathaspids, with separate rostral, pineal and dorsal discs forming the upper part of the armour. Some forms, such as *Lyktaspis* from Spitzbergen, evolved bizarre pointed processes, or rostra, at the front of the armour and widely flared lateral wings. Others, such as *Unarkaspis* from Canada, also had wide lateral spines on the armour as well as high dorsal spines. The common genus *Pteraspis* was recently subdivided into a number of distinct genera by French palaeontologist Alan Blieck. His studies of pteraspids have shown them to be very useful in age determination of Devonian rocks in Europe, Spitzbergen, western Russia and North America. Earlier work by the British palaeontologist Errol White first established a detailed stratigraphic zonation using heterostracan fossils. Some of the well-known pteraspids include *Errivaspis* (named in honour of Errol

systems for sensing their immediate environment. Some amphiaspid fossils have traces of healed bites on them, suggesting that they regularly survived attacks from the larger jawed fishes that inhabited the same shallow seas.

John A. Long

▲ Pteraspid fossils are most commonly represented by parts of the shield, as in this specimen of *Protaspis transversa* from the early Devonian of Wyoming, USA. Shown here is the dorsal surface of the armour. The concentric growth lines can be clearly seen in each of the plates forming the shield.

▶ *Liliaspis* from the Early Devonian of Russia. The tubular mouths of these bizarre-looking amphiaspids were probably used to suck up small prey from the soft muddy sea floor.

Dr Oleg Lebedev, Palaeontological Institute, Moscow

▲ *Tannuaspis*, an unusual heterostracan from the Early Devonian of Russia.

▶ The locality where *Jamoytius* has been found, in the Lower Silurian near Edinburgh, Scotland. Much hard work is required splitting the rock to find a good specimen of *Jamoytius*.

▼ The anaspid *Jamoytius*, from the Silurian of Scotland, was devoid of heavy bony armour and probably lived a lifestyle similar to modern lampreys.

John A. Long

Anatomical features of the anaspid *Jamoytius*.

dorsal fin fold

traces of
body scales

branchial basket
(gill pouches)

anal fin

sclerotic cartilage around eye

mouth cartilage

Dr Alex Ritchie, Australian Museum

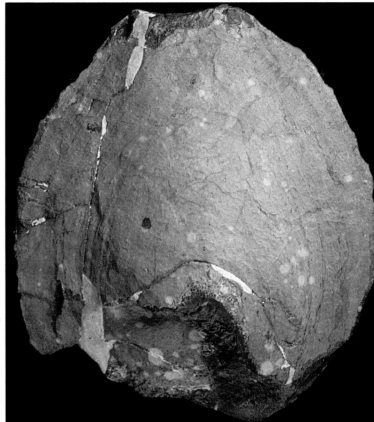

Dr Oleg Lebedev, Palaeontological Institute, Moscow

▲ *Olbiaspis*, a saucer-shaped agnathan from the Early Devonian of Russia.

THE ANASPIDA:
FORERUNNERS OF LAMPREYS

Anaspids were simple, laterally compressed eel-like jawless fishes that may or may not have had a covering of thin elongated scales on the body. They were small, never exceeding 15 cm in length, and flourished during the Silurian and early part of the Devonian. They had simple fins developed along the dorsal and ventral ridges of the body, and some forms, such as *Jamoytius* and *Pharyngolepis,* had well-formed lateral fin folds. The tail was supported by the body axis directed downwards with a thin dorsal (epichordal) lobe. *Birkenia* and *Lasanius* had elaborate arrangements of dorsal scales along the ridge of the body. Like the osteostracans and lampreys, they had a single nasohypophysial opening on the top of the head. The gills opened as a row of holes along the side of the animal, varying in number from 6 to 15 pairs of openings. All known fossil anaspids lived in the ancient Euramerican continent.

Recent discoveries of anaspid-like creatures called *Endeiolepis*, *Euphanerops* and *Legendrelepis* from the Late Devonian of Scaumenac Bay, Canada, lead Drs

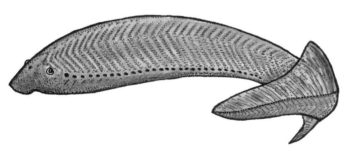

▲ *Birkenia*, an anaspid covered by thin scales, from the Silurian of Scotland.

◄ Reconstruction of *Legendrelepis*, a link between the early fossil anaspids and the living lampreys. (After the work of Dr Philipe Janvier and Dr Marius Arsenault)

▼ *Legendrelepis*, one of the naked lamprey-like anaspids with many pairs of gill openings, from the Late Devonian Escuminac Formation of Quebec, Canada.

Philippe Janvier and Marius Arsenault to suggest that anaspids are the closest fossil ancestors of modern lampreys. The fossil forms from Canada show that they possessed a long row of small circular gill openings that may have stretched almost to the tail, and in *Legendrelepis* this may have been as many as 30 pairs of gills—the highest number of gills in any vertebrate!

The anaspids were probably much like lampreys in their lifestyle, being either parasitic feeders on live fish or detrital feeders. Although their remains have been found largely in freshwater sediments, this does not preclude the possibility that, like lampreys, they may well have spent a phase of their life living in marine conditions.

THE THELODONTI:
THE SCALES TELL THE TALES

The thelodonts (from the Greek, meaning "nipple tooth") are known mostly in the fossil record from their characteristic scales, which have a distinctive "crown" made of shiny dentine on a bony base perforated underneath by a large pulp cavity. Many varieties of scales are known and the tissues making up the scales are also quite variable. Rare whole thelodont fossils show that most were flattened fishes with broad wing-like pectoral fins and large heads with slanting rows of gill openings (for example, *Turinia*). They range in size up to nearly 1 m in length, but most were small fishes, generally less than 15 cm long. Very recent finds of whole, well-preserved thelodonts from the Northwest Territories of Canada show that the group actually radiated into many different forms, some of which had deep bodies with large forked tails and small triangular dorsal fins. The most remarkable feature of these new deep-bodied forms is that they show the presence of a large stomach—an organ thought to be absent in jawless fishes, as it is lacking in living forms such as lampreys.

The oldest fossil thelodonts, known from scales found in Siberia, are of Late Ordovician age, and the group became extinct by the Late Devonian. Most thelodonts had died out by the end of the early part of the Devonian in Euramerica, but those in Gondwana survived later. Their scales range in size from about 0.5 to 2 mm and each thelodont had several thousands of scales covering its body and lining the insides of the mouth and gill slits. This resulted in many different scale shapes for each individual fish; thus the job of sorting out thelodont species from isolated scale populations is really the domain of thelodont specialists.

The thelodonts probably lived a variety of lifestyles. Dr Sue Turner, one of the world's foremost thelodont specialists, believes that the larger, flat forms such as *Turinia* may have been slow-moving

▲ Reconstruction of one of the fork-tailed thelodonts recently discovered in Canada. (After the work of Dr Mark Wilson, Canada)

▼ This is one of the newly discovered fork-tailed thelodonts from the Early Devonian of Canada. These fishes show preservation of a well-formed stomach, an anatomical feature previously thought to be absent in fossil agnathans.

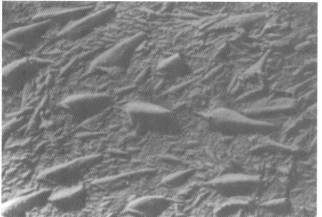

▶▲ *Phlebolepis*, another complete thelodont from the Silurian of Scotland.

▲ Detail of the scales of the thelodont *Lanarkia spinosa*, a good example of how thelodont scales vary within a single individual.

▶ Scanning electron microscope photograph of a turiniid thelodont scale (*Turinia* sp.) from central Australia. Length: 1 mm. Note the dentine crown and bony base.

▼ *Thelodus*, a complete thelodont from the Early Silurian Lesmahagow inlier, near Edinburgh, Scotland.

bottom-feeders, much like the modern angel sharks (*Squatina*), perhaps either grubbing for invertebrates or waiting in ambush for passing prey. The deep-bodied thelodonts from Canada were likely to have been more active swimmers that lived by filter-feeding or catching free-floating prey.

The new finds from Canada support a suggestion that thelodonts were far more advanced than previously thought and may be a closer link to jawed fishes than other agnathans. This is based on the facts that the primitive scales of early thelodonts are very similar to those of early sharks, and that even primitive thelodonts had a well-formed stomach.

Bony shields of galeaspid agnathans are amongst the strangest shapes ever evolved by the vertebrates

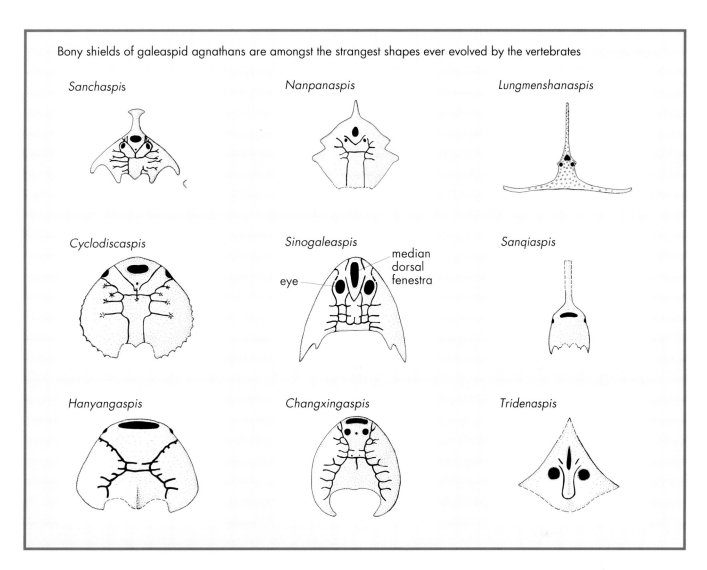

Sanchaspis

Nanpanaspis

Lungmenshanaspis

Cyclodiscaspis

Sinogaleaspis

median dorsal fenestra

eye

Sanqiaspis

Hanyangaspis

Changxingaspis

Tridenaspis

John A. Long

THE GALEASPIDA: MYSTERIOUS ORIENTAL AGNATHANS

The galeaspids are a group of extinct jawless fishes unique to the ancient terranes that today make up China and northern Vietnam. Perhaps because of the long isolation of these ancient continental blocks, the galeaspids evolved a completely different style of armour from other agnathans, some being the most bizarre-looking remains of all fossil fishes. The armour of galeaspids is formed of a single bony shield without separate plates, as in osteostracans and pituriaspids. The unique feature of galeaspids is that the armour has a large median hole in front of the paired eye holes, and this is called the median dorsal fenestra. This is very large in most galeaspids, opening directly below to the paired nasal cavities.

Galeaspids were a diverse group, with more than 40 known species, and the fine preservation of tubes of laminar bone around soft tissues of the head tells us much about their soft anatomy. They had a complex brain and well-developed inner ear with two semicircular canals. The group had evolved by the beginning of the Silurian, and most became extinct by the Middle Devonian, although shields attributed to galeaspids have been found at one site in Ningxia Province, northern China.

The average galeaspids had simple semicircular, ovoid or triangular-shaped shields with well-developed radiating patterns of sensory-line grooves adorning the top surface. Extreme shape development is seen in forms such as *Lungmenshanaspis*, which possessed a drawn-out narrow process of bone at the front and sides of the armour, and others such as *Sanchaspis* which had a forward-projecting tube with a bulbous enlargement on the end. *Polybranchiaspis* was one of the earliest-named galeaspids and has been found throughout China and in the north of Vietnam.

John A. Long

▲ Reconstruction of *Pituriaspis doylei*, named after the hallucinatory Australian plant 'pituri', because of the bizarre shape of the bony armour. It is one of only two known species belonging in the extinct Class Pituriaspida, unique to Australia.

◄ *Polybranchiaspis*, a common galeaspid agnathan from the Early Devonian of China and northern Vietnam.

▼ *Pituriaspis doylei*, a representative of a new class of vertebrates described from central Australia in 1991. This specimen is the cast of the bony shield preserved in Early–Middle Devonian red sandstone. All the original bone has weathered away.

John A. Long

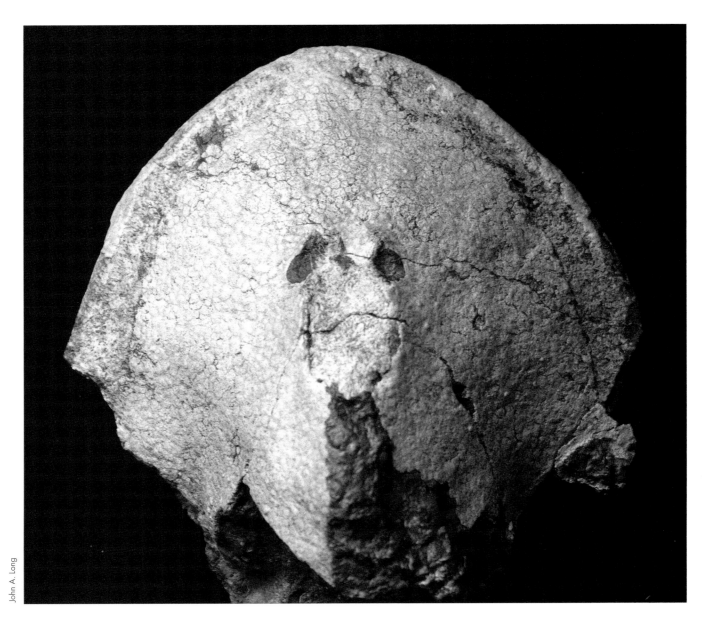

John A. Long

▲ Head shield of the osteostracan *Zenaspis selwayi*, from the Lower Devonian of Wales.

THE PITURIASPIDA: A NEWLY DISCOVERED CLASS OF VERTEBRATES

Pituriaspids take their unusual name from an Australian Aboriginal word *pituri*, the name of a plant containing a narcotic drug sometimes used by central Australian Aboriginals; the fossil was so weird that the discoverer, Dr Gavin Young, thought he may have been hallucinating when he saw it. The pituriaspids are known by two forms, both of which come from the Toko Range in southwestern Queensland, Australia. They represent the only body fossils of agnathans of Devonian age from Australia, and lived close to the start of the Middle Devonian. The unique feature of pituriaspids is that they have a long bony armour, forming a tube around the head and trunk region, and this armour has a large opening below the eye holes. The front of the armour features a long, forward-projecting extension of bone called a rostrum. Two forms are known, *Pituriaspis* and *Neeyambaspis*, the latter having a broader, shorter armour shape. The pituriaspids probably had well-developed pectoral fins, as seen by paired openings on each side of the armour, and a strong shoulder of bone that would have protected the front edge of the fin.

At present we know less about the anatomy of the pituriaspids than other fossil agnathans, but from their general appearance we can place them close to either the galeaspids of China or the osteostracans, all of which have well-developed shoulder processes called the cornual processes. Dr Young classifies the pituriaspids at the base of the radiation containing osteostracans and galeaspids.

THE OSTEOSTRACI: THE PINNACLE OF JAWLESS ACHIEVEMENT

The Osteostraci (once called "ostracoderms") were restricted to the ancient Euramerican continent, and their fossil remains are well-known from the Old Red Sandstone outcrops in Britain, Europe, western Russia, Spitzbergen and North America. They have a large bony shield with two round eye holes, a key-shaped smaller opening for nasal organs, and a tiny pineal opening between the eyes. The sides of the shield have large areas of sensory function, and a similar sensory field sits on the top of the shield. The shield of many species has well-developed spiny processes, called cornua, projecting rearwards. The pectoral fins are well developed in osteostracans, and there may be one or two dorsal fins on the body, as in *Ateleaspis* from Scotland. In fossils the underside of the shield is largely open beneath the cavity for the mouth and gills, but in life was covered by a mosaic of many small platelets. Up to 10 pairs of gill slits were present, as shown by small paired openings beneath the shield. In some osteostracan fossils, impressions of the brain are well preserved and allow scientists to

▲ Close-up of the eyes and pineal region of an osteostracan fish.

► The osteostracan *Hemicyclaspis murchisoni*, from Turin Hill, Wales. The osteostracans were the most advanced of all agnathans in many aspects of their anatomy.

▼ *Tremataspis*, a long-shielded osteostracan from the Late Silurian of the Island of Oesel, near Sweden (length of specimen just under 4 cm).

FISH TEETH

The teeth of vertebrates include a complex diversity of shapes and tissues, and the fundamental building blocks (making complexity possible) first appeared within the evolution of fishes. Although jawless fishes (agnathans) generally lack teeth, some parasitic forms, such as lampreys, actually have tooth-like structures. These are horny, uncalcified, sharp structures around the mouth. As these contain enamel-like antigens and proteins with molecular weights as high as those in gnathostome tooth enamel, they can be argued to be real "teeth". Thus it is quite possible that the origin of teeth predates the origins of jaws. Other fossil agnathans show the presence of pointed bony structures around the mouth, which undoubtedly functioned in a manner similar to teeth although not growing and being replaced as in true gnathostome teeth.

The earliest teeth in fishes belong to acanthodians and actinopterygians dating back to the late Silurian Period. Placoderms of this age also had tooth-like structures on their jaws, but not set into the jaw, and not delineated from the actual jaw bone. Thus placoderm "teeth" are not regarded as true teeth, only "cusps". The osteichthyan fishes developed true teeth comprising layers of enameloid over dentine with a central pulp cavity. The teeth are ankylosed to the jaws, lacking root systems as occurs in chondrichthyan teeth. In most fishes the teeth are homodont—all teeth being similar in size and shape, although variations in size are common in crossopterygians, which have smaller rows of marginal teeth flanking very large stabbing fangs. Fishes that feed by crushing hard-shelled invertebrates may develop pavement-like flat teeth reinforced with tubes of pleuromic dentine, as in holocephalans and bradyodont (cochliodont) sharks.

One of the earliest innovations in tooth development is the "dental lamina" of chondrichthyans, a structure that produces teeth throughout the life of the fish. Sharks thus have an ever-growing supply of teeth, which are periodically shed throughout their life.

The lungfishes include forms with flat crushing tooth-plates, some having numerous rows of teeth on each tooth-plate. Such structures still contain outer enameloid over the individual tooth cusps, which grow out from a dentine base. Hypermineralised tissues such as "petrodentine" strengthen the central region of each cusp on the tooth plate.

Complex labyrinthine infolding of dentine and enamel is a specialised feature that occurs only in certain crossopterygian fishes and early tetrapods (amphibians). It is a means of strengthening the large teeth, and the degree of complexity seen in tooth cross-sections sometimes characterises some of the groups; for example, porolepiforms have "dendrodont" style, others have "polyplocodont" style. The tissue "enamel" (in its strict definition) occurs only in advanced crossopterygian fishes and higher land vertebrates. It is the highly mineralised outer shiny layer of the tooth, which has well-ordered, perpendicular crystallite orientations when viewed in cross-section. Similar kinds of hard outer layers on primitive fish teeth lack the same degree of regular orientation of the crystallites and are termed "enameloid" tissues.

In higher osteichthyans, teeth may occur not only on the jaw bones but also inside the mouth cavity on small dental plates located on the gill arch bones. Certain specialised teleostean actinopterygians take this one stage further by having "pharyngeal mills", pavements of teeth set on hard bony bases in the throat to grind up food.

reconstruct the cranial anatomy. From such fossils we know that they had only two semicircular canals forming the inner ear, as opposed to the three found in higher vertebrates, and that the general plan of the cranial nerves and vascular supply to the head was similar to that in the larvae of lampreys.

The osteostracans underwent a major radiation resulting in a great diversity of shapes, ranging from those with simple semicircular head shields (such as *Cephalaspis*) to others with prominent dorsal spines (*Machairaspis*), or elongated shields that cover much of the trunk of the fish (*Thyestes, Dartmuthia, Nectaspis*). All the fossils that have the tail preserved show the presence of thick scales, often arranged as a

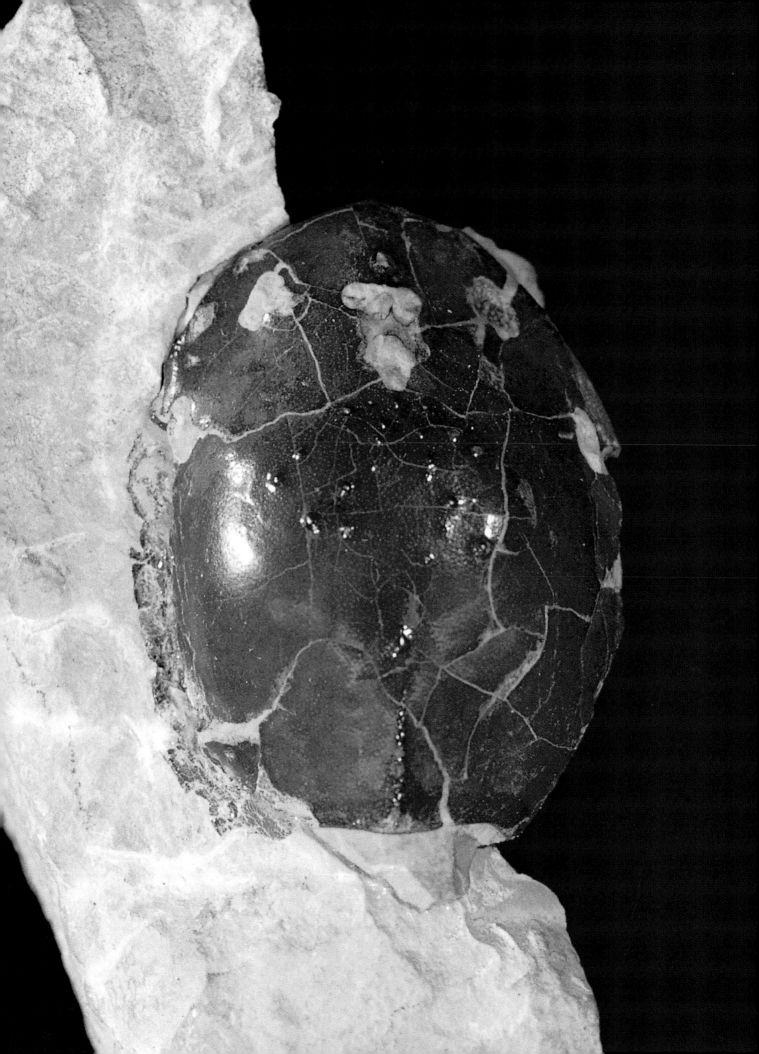

series of vertically oriented rectangular units, which are capped by a series of smaller scales along the back and another series underneath on the belly. The osteostracans died out early in the Late Devonian.

THE END OF THE JAWLESS EMPIRE

By the start of the Late Devonian most of the many families of jawless fishes had become extinct. The handful of survivors included a single species of osteostracan found in Canada (*Alaspis*), one possible galeaspid from Ningxia in northern China, one genus of thelodont from Australia (*Australolepis*), the three lamprey-like anaspids from Canada (*Euphanerops, Endeiolepis, Legendrelepis*) and some of the large, flattened psammosteid heterostracans from Europe. The reason for the rapid decline in agnathan diversity is probably the rapid increase in the diversity of jawed fishes. In the Silurian the agnathans ruled as the dominant fish type, and the early jawed fishes were comparatively rare. By the start of the Devonian all the major groups of jawed fishes had evolved, but by the Middle and Late Devonian many of these groups reached a peak of diversity (such as placoderms, crossopterygians and lungfishes).

It is clear from their similar body shapes that some of these placoderms simply took over the niches of agnathans that succumbed to increasing predation pressure. For example, the flattened phyllolepid placoderms appeared straight after the extinction of the flattened psammosteid agnathans, as we know from their fossils in the same succession in the east Baltic region. Long-shielded heterostracans were probably outcompeted by the long-shielded early placoderms, which could defend themselves better from predators. The detrital bottom-feeding agnathans may have been put out of business by the many new forms of bottom-feeding placoderms such as antiarchs. By the close of the Devonian, 355 million years ago, only the lampreys and the hagfishes, unburdened by bony armours, were left to carry on the jawless tradition.

▲ Reconstruction of an unsual osteostracan, *Machairaspis*, from the Early Devonian of Spitzbergen. (After the work of Dr Philippe Janvier, Paris)

◀ A long-shielded osteostracan, *Dartmuthia*, from the Silurian of Canada (length of specimen just under 5 cm).

LINKS TO THE FIRST JAWED FISHES

The Osteostraci are widely regarded by palaeontologists as the most advanced agnathans on the evolutionary ladder leading to higher vertebrates, as they are the only agnathans that share with jawed fishes the development of paired fins, the open endolymphatic duct on the head, and ossified bones around the eye (sclerotic bones). The vital connection between the presence of eye bones and the origins of jawed vertebrates is seen through recent discoveries in developmental biology. Through studies of the developing chick embryo, Dr Brian Hall, of Halifax University, Canada, discovered that the tissue forming the buds that develop into the sclerotic bones covering the outer margin of the eye are also integral to initiating the formation of the lower jaw. Perhaps it was the developmental pathway for the byproduct of jaw ossification. Once sclerotic bones had formed, a series of connector genes might have accidentally begun making a lower jaw cartilage, perhaps as a means of strengthening the existing mouth parts. Such a model would indicate that tooth-like structures could have evolved well before supporting jaw bones. But is there any evidence for this in the fossil record?

Diverse groups of fossil agnathans show that some did have functioning bony plates in the mouth area for feeding on carrion or even for feeding on live prey. The initiation of an ossified support for these mouth parts, the Meckel's cartilage, may therefore have arisen serendipitously as a byproduct of protecting vital sensory organs, the eyes. The earliest jawed fishes (gnathostomes) are presumed to be the sharks, as their scales date back further than for any other group, to the Late Ordovician–Early Silurian of Mongolia.

The problem of whether these early scales belonged to sharks with teeth and jaws has not yet been resolved, as the occurrence of shark teeth and scales in the same deposits is not known until the Early Devonian. Thus the high degree of anatomical attainment by the osteostracans was to set the stage for the next great revolution in fish evolution, the rise of the jawed fishes. With jaws and teeth the predator–prey arms race was set to begin, and the war for survival or extinction in the fish world was to keep raging throughout the Devonian Period.

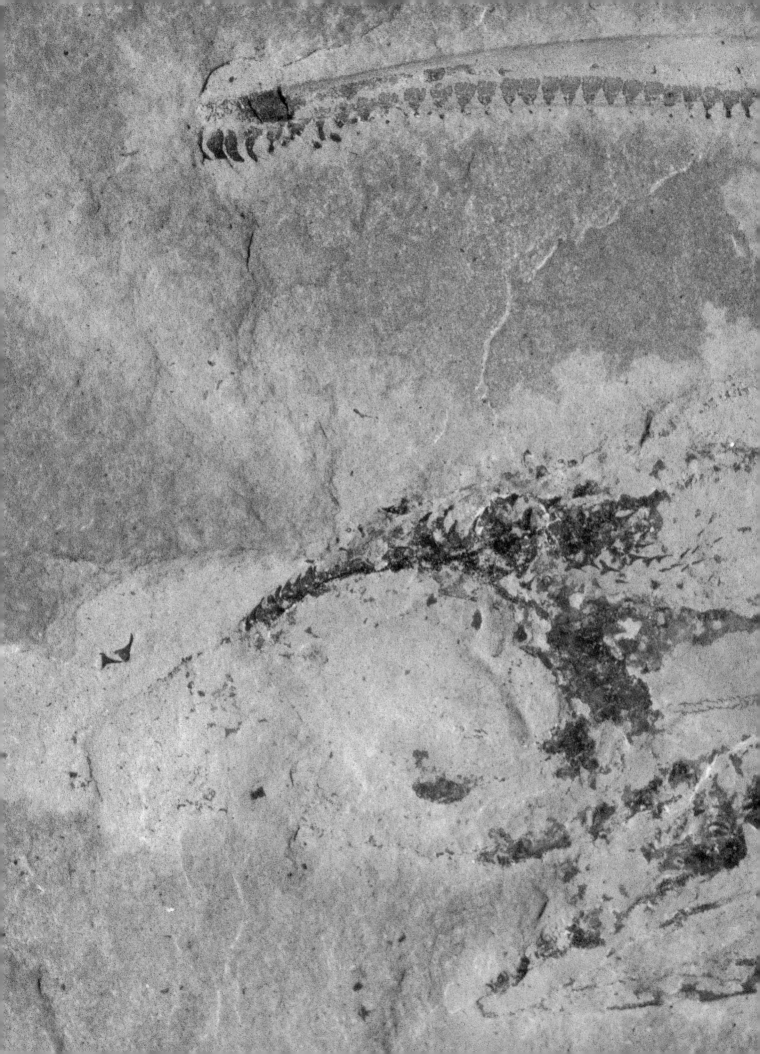

Chapter 3

SHARKS AND THEIR KIN

CLASS CHONDRICHTHYES

Sharks are one of nature's great success stories. Since their first appearance about 400 million years ago, they have changed very little, merely improving their ability to hunt and feed through evolving more resilient tooth stuctures and more streamlined body shapes. The basic body plan of the shark underwent two successful major modifications: one in the Carboniferous Period, when the holocephalans (chimaeras and rabbitfishes) first appeared; and one in the Jurassic Period, when the flattened rays evolved. The acme of chondrichthyan evolution came in the Late Palaeozoic, immediately following the decline of the placoderms. Many bizarre forms of Palaeozoic sharks have been discovered, some with coiled whorls of serrated teeth, others with huge bony structures on their dorsal fins. Today, sharks, rays and holocephalans are known from more than 900 species, including the largest fish on earth, the massive (but gentle) whale shark which grows up to 15 m in length. Huge predators such as the great white shark today reach lengths in excess of 7 m, but only 2 million years ago voracious predatory fossil sharks grew up to 14 m long—the most gruesome underwater death machines that ever lived.

Sharks are perhaps the most feared yet most intriguing of all fishes. Despite the grisly images we might have of efficient killing machines that relentlessly attack humans, humans actually consume shark meat at a rate many millions of times more (by weight) than sharks consume people. Not all sharks are "killers", indeed only a handful of the 344 or more known species have any record of attacking people or would even be capable of inflicting harm. Those that are known to be man-eaters, such as the great white shark, usually feed on seals or large fishes, and may occasionally take a human by accident if the person unwisely chooses to venture into the hunting grounds of the shark. However when we look at sharks from the perspective of their 400-million-year history we see a totally different story, one of a highly successful and efficient predator, which has changed little from its first appearance. Indeed the evolution of sharks is known principally by the fossil record of their teeth and scales and by a few exceptional specimens showing whole body features.

Sharks are among the earliest known jawed fishes, their scales having been found in rocks dating back to the dawn of the Silurian Period some 420 million years ago. However, the fact that shark scales occur at this time without the associated fossilised teeth suggests that these early 'protosharks' may not necessarily have had teeth, or even jaws. The mystery remains as to how sharks first evolved. Did they evolve from a heavily scaled agnathan group, like the thelodonts, or from an as-yet-undiscovered ancestral fish group? The current fossil evidence is too incomplete to answer this question. The only clue we have to their distant origins is the tantalising similarity that exists between the scales of early sharks and those of the jawless thelodonts.

Previous pages: The head of *Damocles serratus*, from the Early Carboniferous Bear Gulch Limestone of Montana, USA. *Dr K. Frickhinger, Munich*

Inset: The great white shark *Carcharodon carcharias* is the largest predatory fish alive today, sometimes exceeding 6 m in length. However, only a few million years ago, giant predatory sharks whose serrated teeth were 18 cm long cruised the world's oceans. These killers may have reached lengths of 15 m. Sharks hit on the right formula from their very beginnings and have changed little from their Devonian ancestors which lived 370 million years ago. *Ron and Valerie Taylor, Sydney*

BASIC STRUCTURE OF PRIMITIVE CHONDRICHTHYANS

Sharks, rays and rabbitfishes belong to the Class Chondrichthyes (meaning "cartilage fishes"), so-named because they lack an internally ossified bony skeleton, having instead a special type of cartilage forming the braincase, jaws, gill arches, vertebrae and fin supports. The only hard bony tissues are developed in their defensive fin-spines, teeth and scales. Although many biologists once thought of the cartilaginous condition of sharks as being primitive, as a precursor to the evolution of true bone, it is now regarded as a highly specialised condition that enables sharks to function more efficiently. The cartilage forming their internal skeleton may be strengthened with needles of the mineral calcite to give a high degree of strength whilst not adding unnecessary weight. The lack of a swim-bladder inside the shark means that buoyancy in the water column is achieved by having a large oil-filled liver in association with the much-lightened skeleton. The broad wing-like pectoral fins give sharks their hydrodynamic lift within the water as long as they keep moving.

Sharks have simple jaws consisting of the primary upper and lower jaw cartilages (Meckel's cartilage and palatoquadrate) armed with rows of teeth that grow throughout life and replace damaged or shed teeth from the front rows. The teeth of sharks and rays are composed of a dentine crown over a bony base pierced by canals. In food-crushing species they may be flattened combs of dentine and bone that form a grinding pavement. Chondrichthyans have placoid scales, which have a bony base and crown made of dentine, and these are set into the skin but do not overlap each other. The scales are simple blade-like structures or complex with several generations of growth on each base.

One of the features that distinguish chondrichthyans from other fishes is that they reproduce by internal fertilisation. The males have intromittent organs, called claspers, attached to the pelvic fins. They insert these into the cloaca of the female to fertilise her eggs. Even the fossil forms show the presence of claspers and many well-preserved sharks from the Early Carboniferous Bear Gulch Limestone of Montana show distinct shape and fin differences between the sexes. Males not only have claspers but also may possess elaborate fin-spines with brush-like structures (as in *Falcatus*, *Damocles* and *Stethacanthus*) which most likely played an important role during mating.

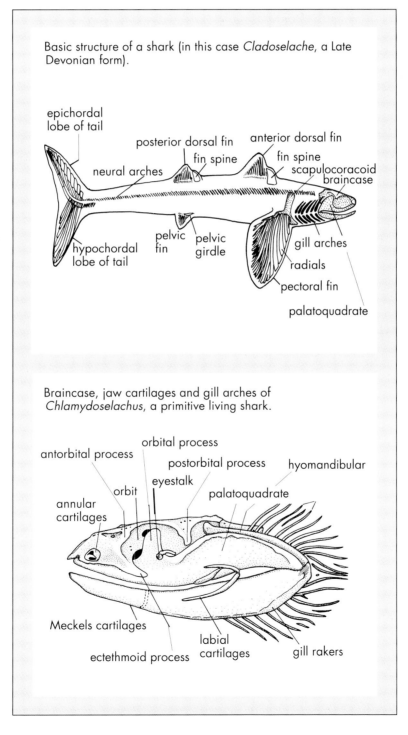

Basic structure of a shark (in this case *Cladoselache*, a Late Devonian form).

epichordal lobe of tail

posterior dorsal fin

anterior dorsal fin

fin spine

neural arches

fin spine

scapulocoracoid

braincase

hypochordal lobe of tail

pelvic fin

pelvic girdle

gill arches

radials

pectoral fin

palatoquadrate

Braincase, jaw cartilages and gill arches of *Chlamydoselachus*, a primitive living shark.

orbital process

antorbital process

postorbital process

hyomandibular

eyestalk

orbit

palatoquadrate

annular cartilages

Meckels cartilages

ectethmoid process

labial cartilages

gill rakers

John A. Long

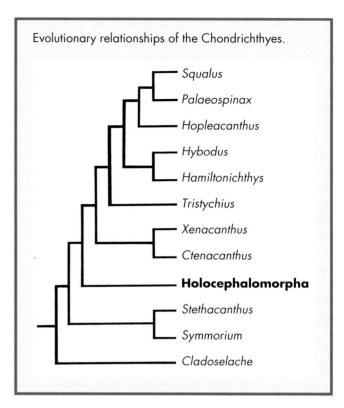

Evolutionary relationships of the Chondrichthyes.

- *Squalus*
- *Palaeospinax*
- *Hopleacanthus*
- *Hybodus*
- *Hamiltonichthys*
- *Tristychius*
- *Xenacanthus*
- *Ctenacanthus*
- **Holocephalomorpha**
- *Stethacanthus*
- *Symmorium*
- *Cladoselache*

▲ The oldest known fossil shark braincase belongs to *Antarctilamna prisca* from the Bunga Beds of New South Wales, Australia, of Middle Devonian age.

▶ Reconstruction of the braincase of *Antarctilamna prisca* showing anatomical features.

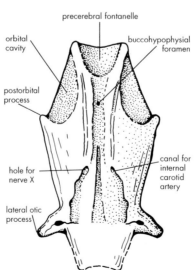

precerebral fontanelle

orbital cavity

buccohypophysial foramen

postorbital process

hole for nerve X

canal for internal carotid artery

lateral otic process

Because sharks are a living group of fishes we have good background knowledge of the anatomy, physiology and lifestyle of many species. Recent research is showing that the shark is far more advanced in many ways than previously thought. Some sharks, for instance, lay eggs and let the young hatch themselves, whereas others give birth to live young and have an equivalent to a mammalian placenta. Some sharks even have quite complex mating behaviour and rituals. All sharks possess extremely sensitive snouts which house numerous pores containing ampullae of Lorenzini. These organs enable sharks to detect sensitive electric fields so that they can find prey that may even be buried below the surface sands of the sea floor. Hammerhead sharks, which have the eyes widely spaced on stalks, are now believed to have evolved such bizarre heads because this enables them to spread their ampullae over a wider area to increase their electrosensitivity.

Sharks live in a great variety of habitats, from near-shore reefs to open oceans; they have been found at all depths, from surface waters to deep abysses; and some species may invade freshwater river systems, travelling hundreds of kilometres inland. It has even been discovered that some sharks, such as the great white *Carcharodon carcharias*, can maintain higher body temperatures than the surrounding seawater, a marvellous physiological advance that enables them to move with great speed and regulate their body temperature irrespective of the changes in the water around them.

The chondrichthyans have a fascinating evolution that is more akin to the long-term fine-tuning of a perfect machine than periodic leaps into new models. From their first appearance, sharks have got it right.

William Stout, Pasadena, USA

▲ Reconstruction of *Antarctilamna prisca*, a Middle Devonian shark from Antarctica and southeastern Australia.

▼ Searching for shark fossils in the Lashly Range, southern Victoria Land, Antarctica. In 1970 the remains of the oldest partially articulated shark, *Antarctilamna*, were found here by Australian scientists Gavin Young and Alex Ritchie. Antarctica and Australia were joined as part of Gondwana in the Devonian, and this region may have been the place where sharks first originated.

John A. Long

▼ A new form of fossil shark's tooth from the Middle Devonian of Antarctica, with two broad cusps on a wide root.

Kristine Brimmell, Western Australian Museum

THE ORIGINS OF SHARKS

The origins of sharks are still a mystery. Some scientists regard sharks as the most primitive of all the jawed fishes, whereas others see them as highly specialised forms that did not require the complex bony ossifications of other fish groups. Sharks appear to be closely related to the now-extinct placoderms (see Chapter 5), and both these groups may have arisen from a scale-covered jawless form well before the Early Silurian. The presence of shark-like scales of this age, and their striking similarity to the scales of the jawless thelodonts, has lead some workers to suggest that thelodonts and sharks could be close relatives, and the recent discovery of the remarkable fork-tailed thelodonts from Canada would seem to support this view.

The earliest "chondrichthyans" are known from scales of the Early Silurian, about 420 million years ago, found in Mongolia. These scales, such as *Mongolepis* and *Polymerolepis*, resemble the simple placoid scales of modern sharks, but because no teeth fossils occur with them in the same rocks we have no hard evidence that these early shark-like fishes actually had teeth or jaws.

▲ Sharks have not changed much in over 360 million years. Reconstruction of *Cladoselache* (above), a Late Devonian shark from North America (length up to 2 m), compared with (right) a modern mandarin dogfish, *Cirrhigaleus barbifer* (length to 1.3 m).

The first occurrence of fossil chondrichthyans known from their distinctive teeth dates back to the beginning of the Early Devonian. These tiny teeth, under 4 mm, are named *Leonodus* (after Léon, Spain) and have been found at several sites in Europe. Other Early Devonian records of fossil shark teeth are from Saudi Arabia, and the earliest diverse assemblage of shark teeth is now known to be from the Middle Devonian Aztec Siltstone of Antarctica.

This connection dawned on me after a gruelling field trip collecting fossil fish remains in Antarctica's South Victoria Land in late 1991 and early 1992. I was searching the Lashly Range where, twenty years before, Drs Gavin Young and Alex Ritchie had found the world's oldest partially articulated shark remains. Gavin Young had discovered a little shark, about 40 cm long, which he named *Antarctilamna* ("lamnid shark from Antarctica"). Impressions of the braincase of *Antarctilamna*, as well as fin spines and teeth, have also been found in Australia and Saudi Arabia. *Antarctilamna* had a robust fin spine preceding the large dorsal fin, and its teeth had two large splayed cusps with smaller median cusps in between. Over the 1991–92 season we found several bone bed horizons in the Aztec Siltstone, which on closer examination were found to be rich in small shark's teeth. All of these occurrences

suggest that the oldest true chondrichthyans—those definitely having teeth of characteristic form and tissue types—came from Gondwana, and that the first great radiation of sharks may have taken place there also.

A peculiar little tooth, first discovered in a Middle Devonian layer of the Transantarctic Mountains of Antarctica, was named *Mcmurdodus* by Dr Errol White of the British Museum (Natural History), in London, in 1968. Recently this type of tooth was found in central Australia in rocks of late Early Devonian age. The teeth are up to 5 mm or so wide and have several sharp, flat cusps arranged almost symmetrically along a broad root. The significance of *Mcmurdodus* arises from the suggestion made by Sue Turner and Gavin Young that its multilayered enameloid crown makes it the earliest known member of the neoselachian group to which all modern sharks belong. The root structure of *Mcmurdodus* has penetrating canals, another feature of modern shark teeth. These new observations on *Mcmurdodus*, along with the many finds of shark's teeth from the Antarctic Devonian collected by me during the 1991–92 field season, and the earliest confirmed shark's teeth occurring in Spain and Arabia (marginal Gondwana countries), suggest to me that this first great radiation of sharks probably took place in Gondwana right at the beginning of the Early Devonian.

By the Late Devonian sharks had become cosmopolitan and more than 30 species are known from around the world. In the United States the Late Devonian black shales of Ohio and Pennsylvania have yielded many fine specimens of complete fossil sharks and isolated shark remains, belonging to approximately two dozen different species, but most commonly found are the species of the well-known genus *Cladoselache*. Some specimens may have been up to 2 m long. *Cladoselache*, with its fusiform body, large wing-like pectoral fins and two triangular dorsal fins, each preceded by a short, stout fin spine, looked quite like many modern sharks. The superb preservation of some Cleveland Shale sharks shows the bands of

▲ *Cladoselache*, from the Late Devonian Cleveland Shale of Ohio, North America, is one of the earliest complete shark fossils. Its large triangular pectoral fins are here clearly shown. Other specimens from this site show muscle bands preserved as well as fossilised remains of the shark's last meal.

▶ Dr Mike Williams, of the Cleveland Museum of Natural History, is here working on an outcrop of the Cleveland Shale, Ohio, North America, where some of the world's finest Devonian shark fossils have been found.

muscles and even the shark's last meal in some cases. *Cladoselache* had five long gill slits, and its powerful jaws have many rows of small multicuspid teeth. This type of tooth, in which there is a central main cusp and several smaller cusps on either side, is generally called the "cladodont" type (after *Cladoselache*) and is adapted for swallowing prey whole, rather than gouging or grasping prey, as used by many modern sharks. Studies by Dr Mike Williams of the Cleveland Museum of Natural History have shown that in 53 fossil sharks with prey preserved in their gut regions, about 64 per cent had their last meal of small ray-finned fishes, about 28 per cent fed on the shrimp-like *Concavicaris*, 9 per cent fed on conodont animals and in one specimen the remains of another shark were found. The orientation of the prey showed that the sharks were catching their prey tail first, then

swallowing the prey whole. Another Late Devonian Cleveland Shale shark, *Diademodus*, looks similar to *Cladoselache* but has distinctive teeth with several equally-sized cusps on each root. Sharks were definitely on the rise as the Devonian drew near its end. They must have been efficient swimmers to avoid being eaten by the giant predatory placoderms, such as *Dunkleosteus*, that also inhabited these seas.

Most of the known Devonian sharks are represented solely by their teeth. The great variety of tooth types shows that they had begun their radiation into adopting many diverse feeding strategies. Whereas primitive sharks tended to have two main cusps on each tooth (for example, *Leonodus*, *Antarctilamna*), Late Devonian sharks predominantly had multiple cusps on each tooth. One extreme example of this is the bizarre teeth of *Siamodus* (meaning "tooth from Siam"), found in Thailand, which has up to eight equally-sized cusps on a strongly arched root.

Other common Late Devonian sharks, known almost exclusively from teeth, are the *Phoebodus* group. The teeth have well-developed roots, which may

Primitive living sharks and their dentitions

The seven-gilled shark *Hexanchus* and its lower jaw tooth

The frilled shark *Chlamydoselachus* and its lower jaw tooth

Teeth of the Late Devonian sharks from near Mae Sariang, northern Thailand (all approximately 1–3 mm long).

Phoebodus

Cladodus

Symmorium

protrude prominently, and the crown has three main cusps, which may or may not be flanked by smaller intermediate cusps. The many species of *Phoebodus* have been reported from around the world in rocks of Middle–Late Devonian age. The later forms (Carboniferous and Permian age) attributed to *Phoebodus* are currently under revision and may actually belong to different genera. One unusual phoebodontid is *Thrinacodus ferox*, first described from Australia by Dr

Susan Turner and now recognised from around the globe. The teeth of this species actually resemble little grappling hooks. Dr Masatoshi Goto, from Japan, believes that the living frilled shark, *Chlamydoselachus*, is actually a survivor from the Devonian Period, as its teeth closely resemble those of phoebodonts. By the close of the Devonian, sharks had become firmly established in both marine and freshwater habitats around the world.

▼ The earliest fossil shark known from its teeth is *Leonodus* from the Early Devonian of Europe. These bicuspid teeth are about 3–4 mm long and come from a small shark estimated to be under half a metre in length. Such finds, together with shark fossils from Antarctica and Australia, suggest that sharks may well have originated in Gondwana.

John A. Long

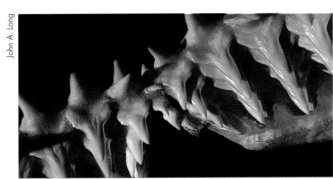

▲ Shark teeth grow in rows from the back and are replaced throughout life, so each shark may shed many hundreds of teeth into the oceanic floor sediments. This is why shark teeth are commonly preserved as fossils.

◄ One of the best Palaeozoic shark fossils ever found is this Carboniferous *Stethacanthus* from the Bearsden site near Edinburgh, found by fossil hunter Stan Wood. The actual specimen is well under a metre in length.

Hunterian Museum, Glasgow

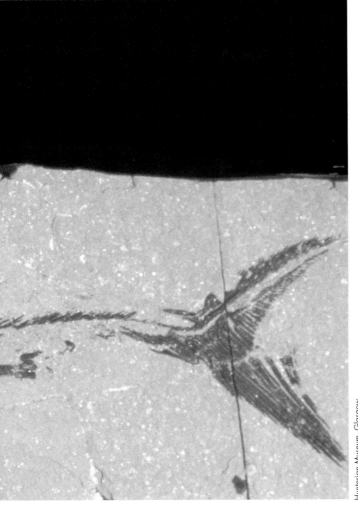

Siamodus

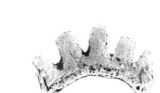

Siamodus

Thrinacodus

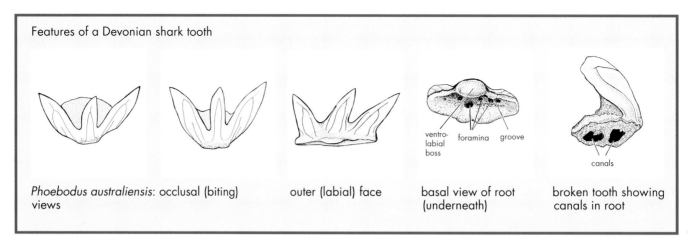

Features of a Devonian shark tooth

Phoebodus australiensis: occlusal (biting) views

outer (labial) face

ventro-labial boss   foramina   groove

basal view of root (underneath)

canals

broken tooth showing canals in root

THE LATE PALAEOZOIC CHONDRICHTHYAN RADIATION

By the Early Carboniferous, following the dramatic extinctions of the armoured placoderm fishes and several other major fish groups, the chondrichthyans underwent another major radiation. Many new families and new higher groups appeared, including the first chimaerids (holocephalans), which are discussed separately below. Rays are also specialised forms of sharks, but they did not arise until the Mesozoic Era.

Some of the Palaeozoic sharks include forms of great variety, much more so than the variations seen in living sharks. The stethacanthid sharks lived in the Late Devonian through to the Permian and had a characteristic bony brush-like structure adorning the main dorsal fin of the males. *Stethacanthus* had a massive dorsal brush, whilst others such as *Damocles* and *Falcatus* had narrow cylindrical dorsal fin spines. A superb whole fossil specimen of *Stethacanthus* was found from the Bearsden deposit near Glasgow by fossil collector

Reconstruction by Mike Coates, Cambridge University

Stan Wood, of Edinburgh, and is now in the Hunterian Museum. Stethacanthids are more commonly known throughout the Late Palaeozoic by their teeth, which have characteristic broad roots with many smaller cusps flanking a principal median cusp. Teeth of Early Carboniferous forms from North America measure up to 70 mm in width, with cusps nearly 4 cm high. This gives an estimated mouth width of nearly 1 m if a regular number of tooth rows (about 12) are assumed to be present, perhaps giving the fish an estimated body length of 6 m or more. Such gigantic hunters prowled the seas as the largest vertebrates in the marine realm of their day, while their predatory counterparts in inland waters were the rhizodontiform fishes (crossopterygians), some 6–7 m long.

Other early sharks of the Carboniferous and Permian Periods include the ctenacanth group, typified by the genus *Ctenacanthus* (from the Greek,

ctenos, "comb", acanthos, "spine"). These sharks had fin spines elaborated by many fine rows of nodes, giving them a comb-like appearance. Complete body fossils of *Goodrichichthys* and *Ctenacanthus* are known from the Early Carboniferous sites near Edinburgh, Scotland. These were generally small sharks, under 50 cm long.

Xenacanth sharks are another group that was successful in the late Palaeozoic and early part of the Mesozoic. Xenacanths have characteristic teeth with two main cusps and a well-developed button of bone on the root (termed the lingual torus). *Xenacanthus*, well known from complete fossils of the Permian–Triassic Periods in western Europe, had a large serrated defensive spine protruding from its neck, and the tail was straight, not heterocercal as in most sharks. Xenacanths were predominantly freshwater predators which invaded river systems from the seas, as their teeth are also well-known from marine deposits. Nearly complete xenacanthid shark remains were discovered quite recently at the Somersby fish site, near Gosford in New South Wales, Australia.

Edestoid sharks were among the most bizarre-looking of any fish. Some of these sharks are known from relatively complete body fossils indicating a streamlined, fast-swimming lifestyle (for example, *Fadenia*). Others, such as *Helicoprion*, evolved complex tooth-whorls that coiled about on themselves and were probably overhung from the lower jaws. There is still much speculation as to how these whorls were used in life. It has even been suggested that they may have mimicked ammonites, coiled shellfish that were abundant at this time, and thus could have attracted prey to the shark if used in a particular fashion. It seems more likely that these sharks used the jagged tooth-whorls when charging into a school of fish or ammonites and thrashing about to snag prey on the projecting array of teeth. Similar feeding methods are used today by the saw sharks. Despite the mystery of how they utilised the tooth-whorls, the group was highly successful, spreading around the world during the middle part of the Permian; *Helicoprion* tooth-whorls have been described from Russia, North America, Japan and Australia.

▲ Reconstruction of the fossil shark *Falcatus falcatus*, exactly as the two individuals were found preserved as fossils. The male is below the female, which is grasping the dorsal spine, present only in the males of stethacanthid sharks, as a prelude to a possible mating ritual. The specimens come from the Early Carboniferous Bear Gulch Limestone, Montana, USA. (After the work of Dr Richard Lund, Adelphi University).

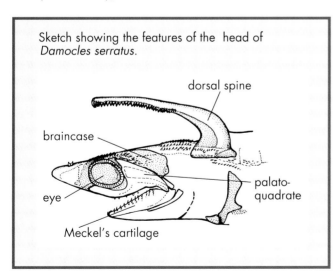

Sketch showing the features of the head of *Damocles serratus*.

dorsal spine

braincase

eye

Meckel's cartilage

palato-quadrate

◀ Reconstruction of *Stethacanthus* showing the peculiar brush of 'teeth' present on the dorsal fin of males. The fin brush may have been used to deter attackers by mimicking a larger tooth-filled mouth when viewed from a head-on approach.

John A. Long

▲ These strange whorls of teeth belong to *Helicoprion*, a Permian shark which had these deadly coils at the front of the lower jaw. They may have used them to thrash about and snag squid-like creatures or other fish on the jagged teeth. This specimen is from the Permian of Russia.

▼ Possible reconstruction of the head of *Helicoprion*, based on closely related edestiod sharks.

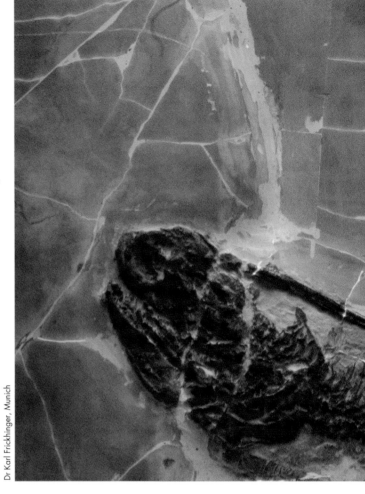

Dr Karl Frickhinger, Munich

▲ *Xenacanthus*, a shark from the Permian of Europe. Xenacanths were large sharks reaching up to 4 m estimated length that stalked river and lake environments from the late Devonian through to the Triassic. They were the top-of-the-line predators in those ecosystems. Some, like *Xenacanthus*, had large serrated neck spines. Their teeth are characteristic in having two large hook-like cusps on each root.

▼ Reconstruction of a xenacanth shark.

William Stout, Pasadena

▲ Close-up of head of the tawny nurse shark, *Nebrius ferrugineus*. The fossil nurse sharks are well known from their teeth, which date from throughout the Tertiary Period.

▼ This strange curved structure is actually a whole lower jaw tooth row of *Pseudodalatias barnstonensis*, a tiny Triassic hybodontid shark. This specimen, under 5 mm in length, was found in the Argillite di riva Di Sotto, Italy.

HUMBLE BEGINNINGS OF MODERN CHONDRICHTHYAN FAUNAS

The great radiation of sharks continued throughout the Mesozoic, and by the end of that era most of the modern shark families had appeared. One particular group was prominent in the early half of the Mesozoic, the hybodontids. These first appear as early as the Carboniferous Period, represented by a beautiful complete fossil of *Hamiltonichthys*, a recently described little shark from Kansas, studied by Dr John Maisey of the American Museum of Natural History. Hybodontid sharks are characterised by their well-developed fin spines and small head spines, as well as by their characteristic broad teeth and scale shapes. The group is best represented by the many species of *Hybodus*, a blunt-headed shark up to 2.5 m long which inhabited the seas of Europe, Africa, Asia and North America in the Triassic, Jurassic and Cretaceous Periods. The teeth of many hybodonts had numerous cusps arranged on a broad root, but some had almost crushing dentitions of flat plated teeth (such as *Acrodus* and *Asteracanthus*). Smaller gripping teeth were set in the front of the mouth.

The seven-gilled sharks (Order Hexanchiformes) appear in the Mesozoic Era, although *Mcmurdodus*, a Middle Devonian form, may possibly also belong in this group. The living cowsharks, *Hexanchus*, *Heptranchias* and *Notorhynchus*, are primitive forms with only one dorsal fin and six to seven pairs of gill slits. Their characteristic teeth have many flat cusps along a broad root, with marked differences between the upper jaw and lower jaw dentitions. They often inhabit deep waters and feed chiefly on fish. Fossil teeth of the group show that forms such as *Hexanchus* have been around since the Early Jurassic.

During the Mesozoic Era many groups of sharks evolved. Some of the commonest fossil teeth found in Cretaceous deposits around the world include the narrow curved teeth of the sand tiger shark, (*Carcharias*) and the broader-toothed lamnids (*Cretolamna*, *Cretoxyrhina*). There are far too many to

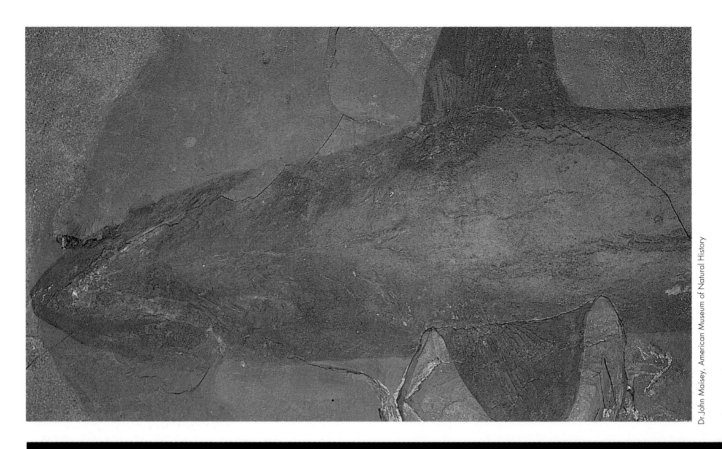

list here, and as far as shark evolution is concerned most of the action was over by this time, as most of the living families had appeared. Beside teeth, other commonly found remains of sharks of this period include their vertebrae, preserved because of the high degree of calcification in the cartilage. Great lamnid sharks appear in the Cretaceous, known from whole skeletons, which show that forms up to 5–6 m long were cruising the seas at the same time as the great reptilian predators, the ichthyosaurs, mosasaurs and plesiosaurs.

▲ Reconstruction of *Hybodus*, a common genus of Mesozoic shark growing to about 2 m long. (After the work of Dr John Maisey, New York)

◀ The head and pectoral fins of *Caseodus*, a primitive edestid shark from the Carboniferous of Nebraska, USA.

▼ *Hamiltonichthys*, the earliest known hybodontid shark, from the Lower Carboniferous of Kansas, USA.

During the Tertiary Period the remaining living genera of sharks evolved, including rare forms like the deep-water goblin shark, *Mitsukurina*, and the megamouth shark, *Megachasma*, and many others that can be traced back through their fossil teeth to their first appearance. The largest predatory sharks of all time evolved by the start of the Miocene, about 23 million years ago. Once thought to be ancestors of the great white shark, these giant killers are now thought to belong in their own extinct family of lamnid sharks. The largest species of fossil mako, *Isurus hastalis*, commonly found in Miocene and Pliocene deposits around the world, may have reached sizes of 6–8 m and, by developing serrations on its teeth, gave rise to the great white shark, *Carcharodon*. Many fossil species of *Isurus* are known from Miocene and younger rocks around the globe, including the teeth of a species alive today, *Isurus oxyrinchus*.

The teeth of the largest predatory shark to have ever lived, *Carcharocles megalodon*, were up to 15 cm long. Although early estimates grossly exaggerated the maximum size of this shark to 25 m or more, it is now possible to constrain the estimates by what we know of how sharks grow and how their teeth increase in size proportionately to overall body length. The revised maximum size of *Carcharocles megalodon* is about 15 m, still twice the size of the largest great white shark ever caught. The characteristic large triangular serrated teeth of this monster are found in rocks of Miocene and Pliocene age all around the world. *Carcharocles megalodon* evolved from an earlier large species, *Carcharocles angustidens*, of Oligocene age. The *Carcharocles* lineage is believed to be descended from *Otodus obliquus*, a

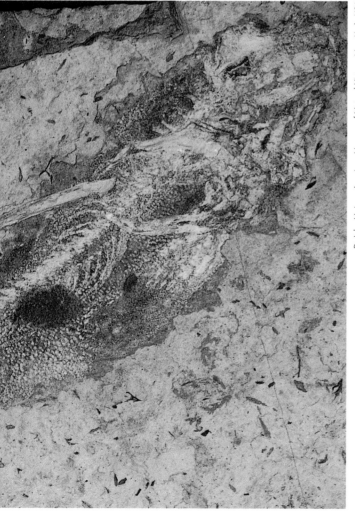

Dr John Maisey, American Museum of Natural History, New York, USA

Dr Alex Ritchie, Australian Museum

▲ Head and trunk of a new fossil xenacanth shark of Triassic age discovered at Somersby, New South Wales, by Dr Alex Ritchie.

John A. Long

▲ Tooth of a Mesozoic hybodontid shark, *Polyacrodus*, showing highly crenulated enamel surface.

▶ This rare complete fossil carcharhiniform shark, *Galeorhinus cuvierii*, is a relative of the living tope shark (*Galeorhinus galeus*). The specimen is Eocene age and comes from the famous Monte Bolca site in Italy.

▼ The Triassic sediments of the Sydney Basin have produced some extremely well-preserved fossil sharks in recent years. The Somersby locality has yielded hundreds of well-preserved fossil fishes.

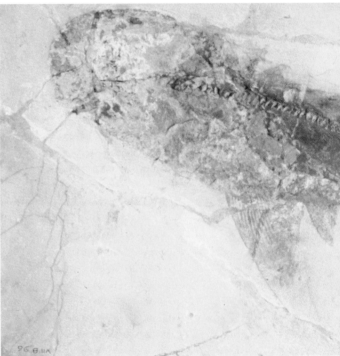

Dr Alex Ritchie, Australian Museum

large form that terrorised the seas during the late Palaeocene–early Eocene. *Otodus* teeth reach sizes of 8 cm or more and indicate that this early large lamnid was up to 6 m long.

It is probably no coincidence that the largest *Carcharocles* species evolved at precisely the same time as the big filter-feeding whales appeared. Reports of giant great white sharks up to 10 m long in recent times have led some scientists to suggest that *Carcharocles* might still be alive out there somewhere, although I regard most fishermen's estimates of "the one that got away" with extreme caution. If the giant *Carcharocles megalodon* was still alive today its teeth would almost certainly have been found in fairly recent marine deposits, but no fossils have been found in strata dating from about 3 million years ago until today. One can only speculate as to why these mighty killers died out. The first drastic climatic cooling that was to herald the Pleistocene ice ages began in the late Pliocene about 2.6 million years ago. Perhaps they could not adjust to the climatic changes as well as their suggested prey, the marine mammals. We know from recent discoveries of fossil baleen whales of this age in Antarctica that the great whales began living in Antarctic waters at this time. Was it just for feeding or also to escape from colder-blooded predators that could not cope with the near-freezing Antarctic waters?

HOLOCEPHALANS: AN EARLY RADIATION

Holocephali is a small subclass known today from about 34 species, such as the elephant shark, chimaerid, and rabbitfish. Like sharks they have a cartilaginous internal skeleton and reproduce using internal fertilisation, the males bearing clasping organs. However, unlike other chondrichthyans, the holocephalans have an operculum covering the gill arches, and their upper jaw is fused to the braincase, providing a powerful bite for their crushing plate dentition. They have long bodies with whip-like tails. They swim by fluttering movements of the pectoral fins and slow sideways movements of the tail.

Before the dramatic discoveries made by Dr Dick Lund of Adelphi University, in the Bear Gulch Limestones of Montana, the early history of holocephalans

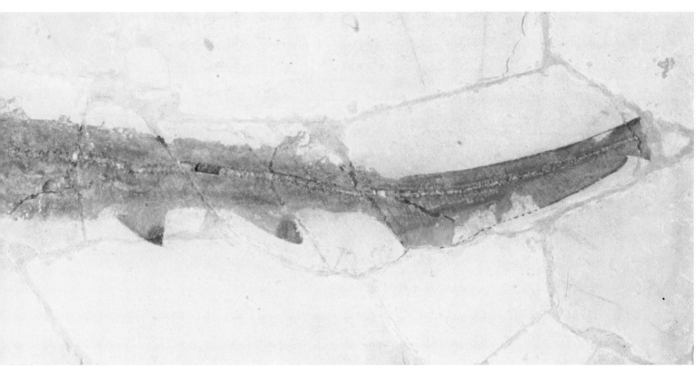

Dr Lorenzo Sorbini, Verona

▼ The blacktip reef shark, *Carcharhinus limbatus*, is one of the carcharhiniform sharks, a group which has been around for nearly 200 million years.

Clay Bryce, Western Australian Museum

was virtually unknown. Early theories had them evolving from the ptyctodontid placoderms, although these were not widely accepted by the palaeontological community. While it was once thought that holocephalans and sharks diverged from a common ancestral form, it is now clear from the fossil record that the holocephalans are an early split off from the mainstream of Devonian sharks.

The Bear Gulch Limestone of Montana has yielded an extraordinary diversity of complete holocephalan fossils; more than 30 new types have been found in the past two decades, and still more are found each field season. One of these, *Echinochimaera meltoni*, has a typical body form much like that of any modern holocephalan. The tail is long and whip-like, with an anal fin close to the tail, and the head is large with big eyes. There is a large bony spine preceding each dorsal fin. The males have a brush-like structure on the first dorsal spine, and several feathery spines over the eyes. Females have a single eye spine and a broader first dorsal fin.

▶ Teeth of the giant lamniform shark *Carcharocles megalodon* are up to 18 cm high, suggesting a maximum size for the fish of about 15 m. This was clearly the largest predatory shark ever to have lived. In addition to its gigantic teeth, its fossilised vertebrae have also been found. Contrary to much of the published literature it is not regarded as an ancestor of today's great white shark *Carcharodon carcharias* but comes from an extinct lamnid lineage. *David Ward*

▼ *Triodus*, from the Permian of Odernheim, Germany.

Other Bear Gulch holocephalans include the strange petalodont and cochliodont sharks. *Belanstea montana* is a well-preserved petalodont recently described by Dr Lund. It had a deep body form, with large feathery pelvic and dorsal fins. The identification of these sharks is largely through the dentition of a few, broad crushing teeth, which indicates a diet of hard-shelled invertebrates. Another Bear Gulch form, *Harpagofututor*, was an elongated holocephalan with large backwards-protruding spines over the eyes of the males. The females were simple in form and almost eel-like in appearance.

Iniopterygians are a peculiar-looking group, some, such as *Iniopteryx* having stout pectoral fin spines projecting high up on the shoulder girdle. In some forms the upper and lower jaws are free, whereas in others the upper jaws are fused to the braincase.

▲ A scene showing fishes in the seas around Antarctica about 40 million years ago, a time when Australia and Antarctica were just separated. On the left are two saw sharks (each about half a metre in length); at centre top is a cod; centre bottom are two angel sharks, flattened ray-like fishes that grew to about 2.6 m long; to the right of the angel sharks are two holocephalans, *Ischyodus*; in the centre of the scene is a school of herring, and silhouetted behind all these is the giant predatory shark *Carcharocles auriculatus* which may have grown to 10 m or more.

Symphysial tooth-whorls are present at the front of the lower jaws. The dentition consists of simple pointed cones or wider teeth with smaller lateral cusplets. Currently some palaeontologists argue that the iniopterygians are highly specialised forms of holocephalans, whereas others prefer to classify them as their own unique group of chondrichthyan.

Chimaerids were bottom-dwelling fishes that fed chiefly on shellfish and crustaceans. They first appear in the Carboniferous Period and emerged as a successful group in the Mesozoic Era, reaching a peak of diversity in the early Cretaceous. Fossil chimaerids such as *Edaphodon* and *Ischyodus* are known throughout the world from their characteristic crushing tooth-plates. Some of these belong to fishes of large size, up to 3 m or more in length. Today the living chimaerids are principally small deep-water forms, and in some countries such as New Zealand are commercially fished and sold as "flake".

▲ "Jaws" eat your heart out! A smallish *Carcharocles megalodon* is here shown reconstructed with a 2 m human for scale. *Carcharocles* became extinct at least 2 million years ago, so has never been a threat to divers.

▼ Head of the extant holocephalan *Callorhynchus*. These are deep-water chondrichthyans with crushing tooth-plates and an operculum covering the gill chamber. The group can be traced back in geological time to origins about 340 million years ago.

Dr Karl Frickhinger, Munich

John A. Long

▲ *Echinochimaera meltoni*, one of the oldest known holocephalans, comes from the Early Carboniferous Bear Gulch Limestone, Montana, USA.

▼ This superb specimen of the holocephalan *Deltoptychius* comes from the Lower Carboniferous marine sediments near Edinburgh. *Deltoptychius* was a flattened, bottom-dwelling chondrichthyan with a fused plate of bone protecting the head region. Its crushing tooth-plates were probably used to feed on shellfish.

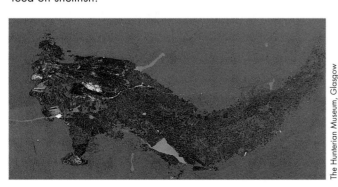

▲ *Belantsea montana* from the Bear Gulch Limestone, Montana, USA, highlights the bizarre body shapes evolved by the earliest holocephalans.

Dr Karl Frickhinger, Munich

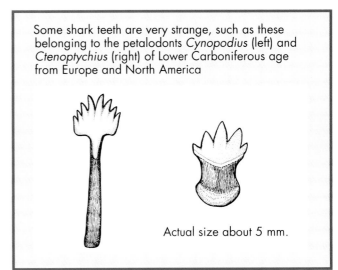

Some shark teeth are very strange, such as these belonging to the petalodonts *Cynopodius* (left) and *Ctenoptychius* (right) of Lower Carboniferous age from Europe and North America

Actual size about 5 mm.

THE RAYS SHINE THROUGH

The first fossil rays date from the Jurassic Period as more or less flattened sharks, which had enlarged their pectoral fins and reduced the tail. Many types of sharks specialised to adapt to a bottom-feeding lifestyle, and living species of angel shark (genus *Squatina*) show well this intermediate shape between the active swimming shark and the flattened ray. The main difference between sharks and rays is that true rays swim entirely by wave-like motions of their wings or pectoral fins, whereas sharks are propelled solely by their tails. Other features seen in rays that distinguish them from typical sharks are the downwards-facing mouth with rows of flat crushing tooth-combs and a well-developed tail

spine (actually a modified dorsal fin spine).

Despite their conservative body plan, rays have evolved to make use of numerous feeding niches. Some giants such as the manta ray are filter-feeders like their shark counterparts, the whale shark and basking shark. Others such as the electric rays have developed the ability to project powerful electric fields to stun the fishes on which they feed or to deter attackers. Many rays simply feed on crustaceans and shellfish by cruising along the sea floor detecting food with their keen electrosensory systems.

The oldest fossil rays are known from the Early Jurassic, represented mostly by isolated teeth and spines. *Spathobatis*, known from the Early–Late Jurassic, is known from well-preserved body fossils of Late Jurassic age in France. *Belmnobatis*, also from France, and *Asterodermus*, from Germany, are other well-preserved rays of similar age. All of these early rays belong to the Family Rhinobatidae which includes several

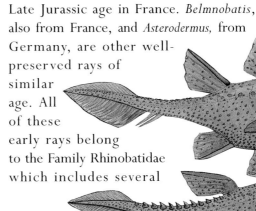

▲ Reconstruction of the female (above) and male (below) of *Echinochimaera meltoni*. (After the work of Dr Richard Lund)

◀ Reconstruction of the unusual petalodont *Belantsea montana*. (After the work of Dr Richard Lund)

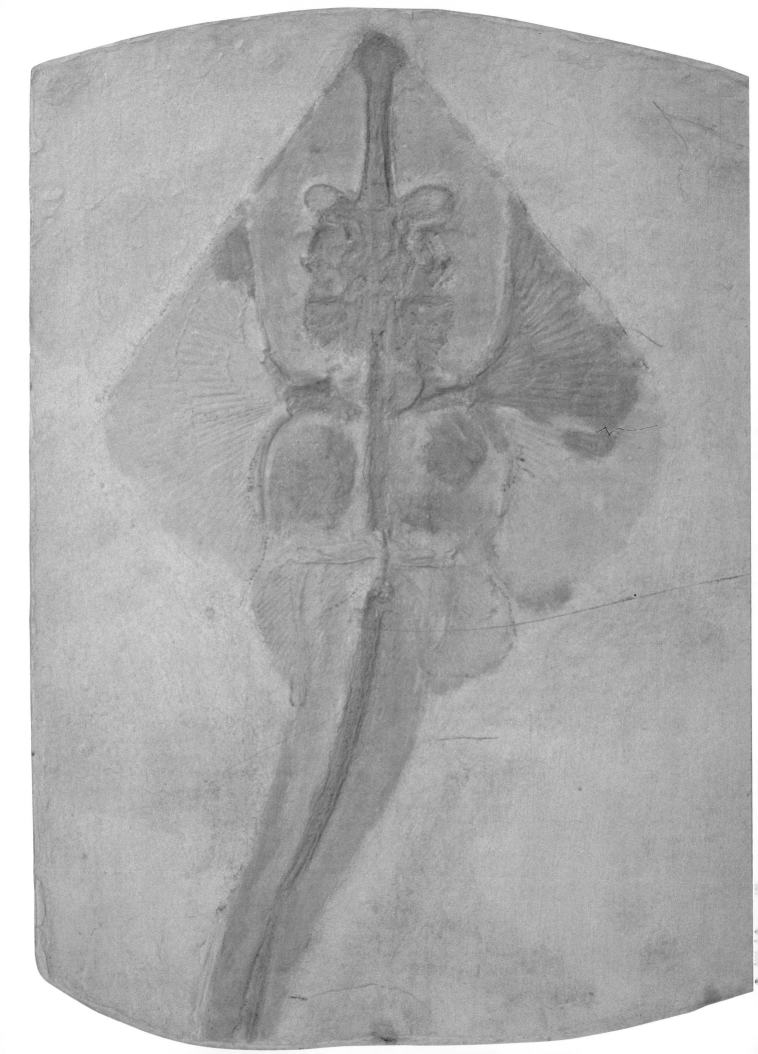

▲ The white-spotted shovelnose ray *Rhynchobatis djiddensis* is one of the primitive rays that are more shark-like in their elongated body shape. The earliest fossil rays are of Jurassic age, and belonged to this same family, the Rhynchobatidae.

▶ This superb fossil stingray *Heliobatis* comes from the Eocene Green River Shales of Wyoming, USA. *Dr Lance Grande, Field Museum, Chicago*

▼ A lower jaw (mandibular) tooth-plate of the extinct chimaerid fish *Ischyodus* from the Eocene of England. *David Ward*

The fossil crushing teeth or 'combs' of the eagle ray *Myliobatis* are commonly found in Miocene to recent marine deposits around the world. *David Ward/David Kemp*

◀ Modern stingrays resembled the fossil forms (right from their first appearance), such as this Jurassic rhinobatid from Solnhofen, Germany. Most of the fossil rays have closely-related living relatives.

▶ *Iniopteryx*, a bizarre fossil chondrichthyan from the late Palaeozoic of North America believed to be allied to the holocephalans.

living forms, such as the shovel-nose ray and the guitarfish. These all have relatively long, shark-like bodies.

The Family Myliobati-dae, or eagle rays, made their first appearance in the Late Cretaceous, rapidly be-came widespread throughout the Tertiary and are still thriving today. The eagle rays have complex pavement-type dentitions consisting of several rows of broad, flat crushing plates flanked by many smaller polygonal units. They use these to grind up food such as crabs and shells. *Myliobatis* and the many other known families of rays are commonly found in Tertiary marine deposits around the world, repres-ented by these distinctive crushing plates as well as occasional fossil stingers (modified dorsal fin spines) and scales.

The rays began invading river and lake habitats early in the Tertiary, as seen by the well-known com-plete body fossils of *Heliobatis* from the famous Early Eocene Green River Shales of Wyoming. *Heliobatis* probably invaded the lake systems when the sea levels rose high enough for saltwater to flood the lake system, and so the Green River stingrays are not considered to be primary freshwater fishes.

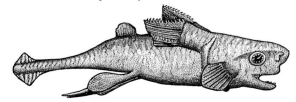

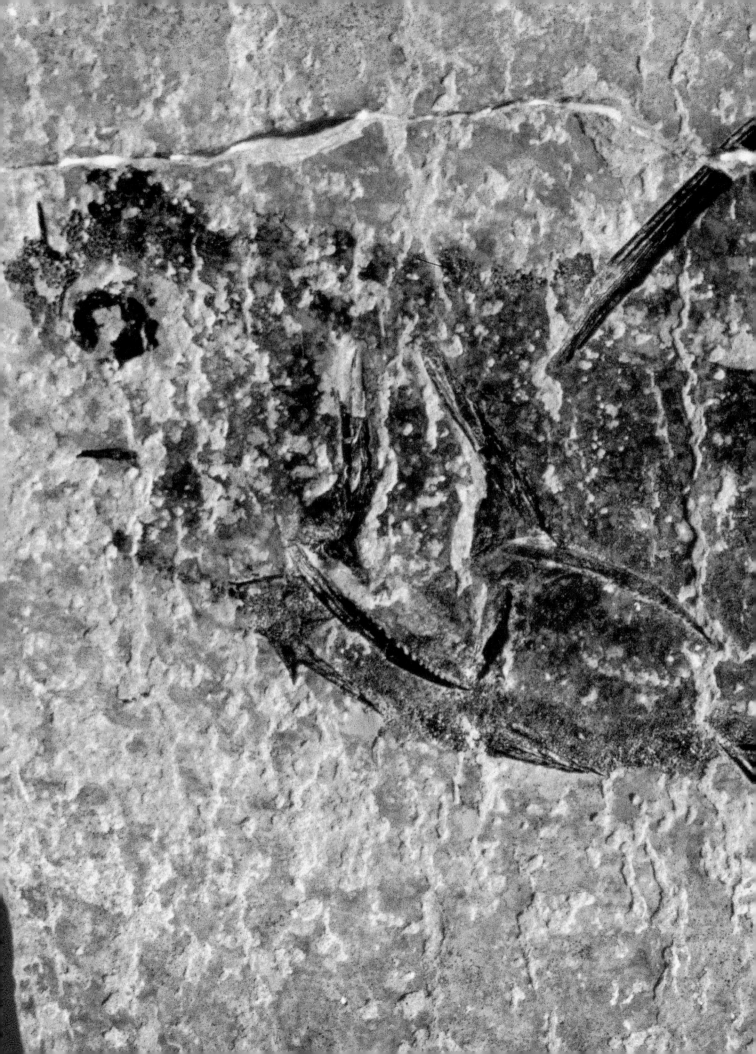

Chapter 4

EXTINCT SPINY FINS

THE ACANTHODII

The acanthodians were an unusual group of mostly small fishes that appeared in the Silurian Period and reached a peak of diversity in the Devonian Period. They are unique among fishes in having ornamented bony spines in front of all fins, and tiny scales that have bulbous, swollen bases. There were three main types of acanthodian: the Climatiiformes, the Ischnacanthiformes and the Acanthodiformes, each distinguished by the presence or absence of bony armour bracing the pectoral fins, the presence of one or two dorsal fins, and the nature of the teeth and scale types. The earliest forms, the climatiiform acanthodians, had elaborate bony shoulder girdle armour and numerous sharp spines. The ischnacanthiforms were predators with teeth firmly fused to the jaw bones. The longest surviving group of acanthodians were the acanthodiforms. These were fast-swimming filter-feeders that lacked teeth in the jaws and had gill rakers for straining food through the gill chamber. The eventual demise of the acanthodian fishes at the close of the Permian Period probably came about through increased competition for food, or predation pressures, brought on by the rapidly growing numbers of ray-finned fishes and sharks.

The first acanthodians were described by Swiss palaeontologist Louis Agassiz in 1844, who named *Acanthodes*, *Cheiracanthus* and *Diplacanthus*. They take their name from the genus *Acanthodes* (from the Greek, *akanthos* meaning "spine") as the group's most distinguishing features are the presence of paired fin spines in front of the pectoral and pelvic fins, and single fin spines preceding the dorsal and anal fins. In many books they have been dubbed the "spiny sharks" although they now appear to be closer to true bony fishes than to sharks. It is their distinctive spines, and the characteristic shape of the acanthodian scale, that makes the group widely recognised from every continent in sedimentary rocks of Silurian–Permian age. Although the oldest acanthodians date back to the beginning of the Silurian, the earliest complete acanthodian fossils are from the Lower Devonian rocks of Europe and Canada.

Acanthodians remain as one of the most enigmatic of all fish groups, about which we have the least amount of anatomical knowledge and few real clues as to their affinities with other fish groups. Unlike the superb Gogo placoderms and other groups of fossil fishes, we lack any acid-prepared, three-dimensional

▲ The acanthodian fishes were alive from the Silurian to the Permian Period, nearly 200 million years. Despite this time span they are a poorly known group, with few known fossil specimens that show details of their internal features. Only the last and perhaps most specialised member, *Acanthodes*, has good preservation of the braincase, which in most other forms may have been cartilaginous.

Previous page: Diplacanthus horridus, one of the advanced climatiform acanthodians from the Late Devonian Escuminac Formation, Canada. *Dr Marius Arsenault, Parc de Miguashua*
Inset: Acanthodes bronni, from the Permian of Lebach, Germany, was the last acanthodian species, and one of the best preserved. This cast shows details of the jaws and gill arches. The Class Acanthodii became extinct by the close of the Palaeozoic Era. *John A. Long*

skulls of acanthodians to study the detailed anatomy of the braincase and reconstruct the soft tissues of the head. Instead, our knowledge of the group is known largely from the shape of the body and fins, the preservation of some head structures in early forms, and the well-ossified head and gill arch skeleton in the last and most specialised species, *Acanthodes bronni*.

In recent years there has been a renewal of studies on acanthodian fossils because their minute scales are useful in determining the relative age of sedimentary rocks. However, the study of their anatomy—for solving the mystery of their evolutionary origins and relationships— has largely been put on hold until new discoveries are made. Several of the newest species of acanthodians, described in the past decade or so, have come from Australia. The first genus of fossil fish I ever described was an acanthodian which I named *Culmacanthus* (from the Aboriginal word *culma* meaning "a spiny fish"). At that time (in 1983) I was a student working principally on osteich-thyan fossils for my doctorate and the unusual fossil remains of the deep-bodied fish (the first specimen of which lacked the characteristic fin-spines) really threw me off the track. At first I thought I had discovered a missing link between acanthodians and osteichthyan fishes as the single specimen I had showed three large ornamented plates, much like the three extrascapular bones at the back of the skull in crossopterygians. I began writing what I thought would be a landmark paper for the prestigious journal *Nature*, and sent a draft copy to British palaeontologist Roger Miles. My rationale was that if a missing link between acanthodians and osteichthyans existed, it probably had not been found previously because no-one had searched the East Gondwanan region for such a beast. I had imagined this fish would be part acanthodian (with tiny scales) and part osteichthyan (with recognisable large head bones). Luckily for me, I soon found some other remains of the same fish, this time showing the characteristic fin-spines of an acanthodian, but so different from any others that it constituted a new genus. I'll never forget the feeling of discovery when, as a student, I recognised my first new genus, which was then placed in a new family, the Culmacanthidae. To me, acanthodians are still a very special and highly interesting group of fishes.

THE ORIGINS AND AFFINITIES OF ACANTHODIANS

The oldest acanthodian fossils, from the Early Silurian of China, are broad fin spines of fishes named *Sinacanthus* (meaning "Chinese spine"). Early acanthodian scales show a simple tissue structure but nonetheless still grew continuously throughout life, a condition not seen in fossil sharks, and a strong link with the higher jawed fishes such as placoderms and osteichthyans. The structure of the head and the shape of the body in the earliest complete acanthodian fossils tell us little about which other group of fishes they may have evolved from or collaterally with. The arrangement of scales in many rows per vertebral segment (micromeric squamation) that typifies acanthodians is otherwise seen in fossil sharks, in some

◄ Reconstruction of *Climatius reticulatus*, one of the early heavily-armoured climatiiforms that lived in the Early Devonian rivers of Britain.

placoderms, and in some actinopterygians, including the most primitive known member of that group, *Cheirolepis* (see Chapter 7). None of the potential ancestral groups of agnathans have similar micromeric squamation. Acanthodians appear to share more specialised characteristics with osteichthyans than with other groups—such as the presence of a similar-shaped braincase, the presence of branchiostegal rays, the nature of the hemibranchs (gill filaments) along the gill arch, and the possible presence of endochondral bone in the braincase (a controversial characteristic, but one advocated by me). Their shark-like features would seem to be mostly primitive characteristics seen in many early jawed fishes, and not suggestive of a close affinity with sharks. This viewpoint, however, is not held by all, as Professor Erik Jarvik, of Stockholm, would argue that acanthodians are actually close relatives of the primitive seven-gilled sharks.

THE CLIMATIIFORMES: ARMOURED AND ADORNED ACANTHODIANS

The oldest group of acanthodians are the Climatiiformes, named after a genus (*Climatius*) found in the Early Devonian Old Red Sandstone of Britain. They are characterised by elaborate bony

BASIC STRUCTURE OF ACANTHODIANS

The acanthodians are nearly always slender elongated fishes with long heterocercal tails and short, blunt heads. The mouth is nearly always large, except in a few specialised forms (such as diplacanthids) and the head is usually invested with many small dermal platelets called tesserae. There are two dorsal fins in all acanthodians but the Acanthodiformes, which have only one dorsal fin situated close to the tail. The eyes have sclerotic bones of varying number encircling them, and there are two pairs of nostrils (excurrent and incurrent nares) at the front of the head. The jaws are ossified as a single lower jaw cartilage which may be supported by a strip of external, ornamented bone (the mandibular splint), but externally ornamented tooth-bearing bones are not present. There are five gill slits opening at the side of the head and these may have dermal branchiostegal plates preceding them in some species, a feature seen otherwise only in the bony fishes.

The gill arch bones and braincase are known only in *Acanthodes*, one of the most specialised of all acanthodians. The gill arches feature a series of basibranchial bones, large hyophyals and ceratohyals, large epihyal and epibranchials, and short, rearward-facing pharyngobranchials. The upper jaw (palatoquadrate) may have a simple articulation with the braincase or have a complex double articulation (in some acanthodiforms).

The braincase in *Acanthodes* is incompletely ossified, made up of four bones that were held together by cartilage. The large dorsal ossification covered most of the top of the head and protected the brain, and a smaller occipital ossification was present at the rear of the head, serving as a site for trunk muscle attachment. The underneath of the braincase had a large anterior basal ossification pierced by a canal for the hypophysis—a space wherethe pituitary gland is housed and the internal carotid arteries converge from outside the braincase, and a smaller rear section below the occipital ossification. In general shape and proportions the braincase of *Acanthodes* resembles that of a primitive ray-finned fish. In some acanthodians the inner ear canals are preserved, indicating that three semicircular canals were present, and furthermore that otoliths, or ossified ear stones, were present in the inner ear of some species (for example, *Carycinacanthus*). These gave the fishes an improved sense of balance and direction, necessary for fast turns and manoeuvres.

The body has a shagreen of many tiny close-fitting scales, and these have a swollen base that lacks a pulp cavity. The scales grew by concentric addition of new layers like an onion. Two main scale types are recognised by their histology: an *Acanthodes* type, which has the crown made of true dentine and a thick acellular bone base; and a *Nostolepis* type scale which has a dentine crown penetrated by vascular canals and a base of cellular bone. The crowns of acanthodian scales are generally quite flat or weakly domed and are separated from the base by a well-defined "neck". Ischnacanthid acanthodians have well-developed jaw bones with teeth and may develop complex dentitions with tooth "fields" present. Climatiid acanthodians had spiny individual teeth and complex tooth-whorls at the front of the lower jaws, whereas acanthodiforms were toothless filter-feeders.

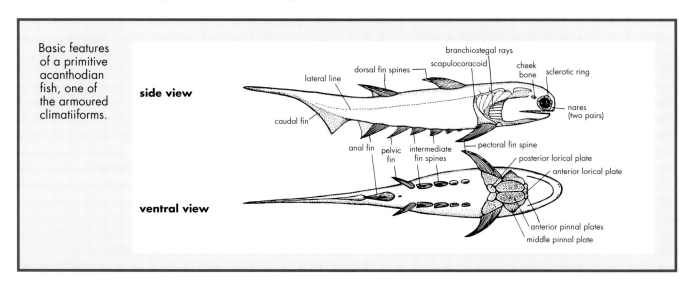

Basic features of a primitive acanthodian fish, one of the armoured climatiiforms.

side view

ventral view

lateral line
dorsal fin spines
branchiostegal rays
scapulocoracoid
cheek bone
sclerotic ring
nares (two pairs)
caudal fin
anal fin
pelvic fin
intermediate fin spines
pectoral fin spine
posterior lorical plate
anterior lorical plate
anterior pinnal plates
middle pinnal plate

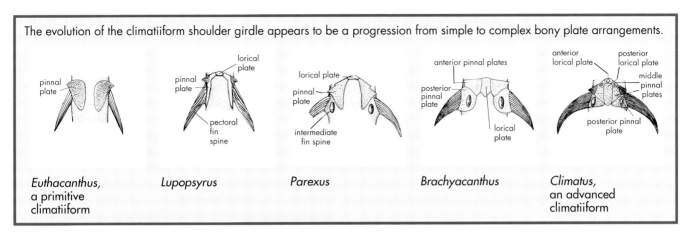

The evolution of the climatiiform shoulder girdle appears to be a progression from simple to complex bony plate arrangements.

Euthacanthus,
a primitive
climatiiform

Lupopsyrus

Parexus

Brachyacanthus

Climatus,
an advanced
climatiiform

▲ An acanthodian from the Delorme Formation, Northwest Territories, Canada. The large defensive spines on the back are reminiscent of the British form *Parexus.*

▶ *Parexus,* an Early Devonian climatiid with large defensive dorsal spines found in Britain.

armour around the entire shoulder girdle, and in primitive forms there may be numerous additional intermediate fin spines in paired rows along the belly, between the pectoral and pelvic fins. The scales and some isolated fin-spines of *Climatius* and its relatives have been recognised from sites in Europe and Russia, and climatiid-like scales have been found almost everywhere in sequences of similar age. Such heavily armoured fishes weren't able to move their pectoral fins, which were rigidly fixed to the shoulder girdle armour, and so the fish probably hydroplaned its way along the sea floor searching for prey. The many sharp spines projecting at all angles from its body were no doubt a deterrent to would-be attackers.

Other heavily armoured early climatiids include *Parexus,* which had one very large spine, and *Eutha-canthus,* which had five pairs of intermediate fin-spines along its belly surface. Many of these fishes come from sites in Britain, which were freshwater/marginal marine deposits. The earliest acanthodians were principally marine fishes, but later in the Devonian they became almost exclusively freshwater forms.

After the first radiation of heavily armoured climatiids had waned (by the Middle Devonian), the group continued in a less spectacular form. The dipla-canthids were a deep-bodied group of climatiiforms, which had greatly reduced their pectoral fin armour, and they had unusual cheek plates covering the side of the head. The best known form, *Diplacanthus* (meaning "paired spines") has been found in Middle–Late Devonian rocks in Europe and North America. *Diplacanthus* retained some primitive features, such as intermediate fin spines and fairly rigid pectoral fins. *Rhadinacanthus* looks similar to *Diplacanthus* but has shorter anterior dorsal fin spines than the posterior dorsal fin. Another diplacanthoid known only from East Gondwanan regions is *Culmacanthus,* a deep-bodied fish which had fewer plates around the pectoral fins and lacked intermediate fin spines entirely. Its characteristically ornamented cheek plates were large and have been found at a number of sites throughout Australia and Antarctica, and the genus is now known from three species, based on the shapes of the cheek plates. The diplacanthids all had small mouths that were devoid of teeth, suggesting that they were algal-grazers or detrital-feeders.

The last climatiiforms were the gyracanthids, large acanthodian fishes up to about 1.5 m long, which first appeared in the Middle Devonian of Antarctica (part of East Gondwana) and radiated out to invade the northern hemisphere countries during the Carboniferous Period. The characteristic large spines with chevron-type radiating ridges are easily recognised in Carboniferous red beds around the world. The only articulated specimens showing what these fishes were really like belong to *Gyracanthides murrayi* found near Mansfield, in southeastern Australia. In this species the shoulder girdle has been freed from the rigid bony plates locking the fins, so that the large pectoral spines were actually capable of a slight degree of movement. These spines were enormous, nearly half the length of the fish, and probably acted as a defence against the large predatory rhizodont fishes that shared its habitat. A closely related form, *Gyracanthus*, is commonly found as isolated spines and shoulder bones in the British coal measures of similar age.

John A. Long

▶ *Nostolepis*, a widespread ischnacanthiform known throughout the world from its characteristic scales found in Late Siluriuan and Early Devonian rocks. This specimen is about 1 mm in length.

◀ The cheekplates, scales and fin spines of the acanthodian *Culmacanthus stewarti*. Note the distinctive ornament on these plates.

▼ Reconstruction of a school of *Culmacanthus antarcticus* swimming in a Late Devonian lake.

John A. Long

William Stout, Pasadena

◀ *Diplacanthus striatus* a Late Devonian diplacanthid acanthodian from the Escuminac Formation, Quebec, Canada.

▶ Reconstruction of *Gyracanthides murrayi* from the Lower Carboniferous red sandstones near Mansfield, Australia.

▼ The unusual deep-bodied acanthodian *Culmacanthus stewarti* from the Late Devonian of Victoria, Australia.

John A. Long

ISCHNACANTHIFORMES: THE PREDATORS

The special feature that distinguishes the Ischnacanthiformes (from the Greek meaning "thin spines") is that they have robust dermal jaw bones with rows of teeth fused firmly to them. These special "gnathal bones" line only the biting margin of the jaws, the rest being largely cartilaginous or, in some species, with an additional bony splint on the outside of the lower jaw. Although ischnacanthid scales are well known from several Middle–Late Silurian sites around the world, the earliest complete fossils of the group are of Early Devonian age, from Canada and Britain. The superbly preserved specimens from the Delorme Formation of Canada's Northwest Territories show that the first ischnacanthids lacked the bony armour around the shoulder girdle that characterised their cousins, the climatiiforms.

One of the best known forms, *Ischnacanthus*, was a slender-bodied predator with long, narrow fin spines. The distinctive gnathal bones show a series of large triangular, curved teeth with smaller cusps separating the large teeth. The head has no large bony plates, only some slightly larger bones in the region of the jaw

joint. One of the oldest and most primitive ischnacanthids is *Uraniacanthus*, from England, which has intermediate fin spines present between the pectoral and pelvic fins.

The largest of all the acanthodians was the ischnacanthid *Xylacanthus*, which, like many ischnacanthids, is known only from its jaw bones. *Xylacanthus* jaws have been found in the Early Devonian red beds of Spitzbergen, indicating a maximum estimated length of about 2 m, which makes this fish one of the larger predators in the shallow marine environments it inhabited. Ischnacanthid jaw elements have been described from many other Devonian marine deposits, some having highly complex dentitions. *Rockycampacanthus* and *Taemasacanthus*, from the Early Devonian limestones near Buchan and Taemas, in southeastern Australia, were small fishes whose jaw bones have a second row of teeth in addition to the main cutting row, as well as having isolated scattered teeth and denticles on the jaws. Overall the ischnacanthids are a poorly known group, and in the few cases in which they are well preserved as whole body fossils, they appear to be very conservative throughout their evolution. Most ischnacanthids died out by the close of the Devonian, although there are some controversial possible exceptions, such as *Acanthodopsis*.

Acanthodopsis, known from jaws with simple triangular teeth, flourished in the Carboniferous coal swamps of Britain, and may have grown to up to 1 m long. However as this form lacks separate gnathal bones with the distinctive, complex tooth morphology seen in other ischnacanthids, in my opinion *Acanthodopsis* may be an aberrant acanthodiform, turned from a primarily filter-feeding lifestyle to predation. Until *Acanthodopsis* is studied in more detail, its evolutionary position will remain in dispute.

THE ACANTHODIFORMES: FILTER-FEEDERS AND SURVIVORS

The Acanthodiformes were the most successful group of acanthodian fishes and are easily characterised by their single dorsal fin, lack of teeth, and well-developed gill rakers. These features indicate a free-swimming, filter-feeding lifestyle. A major trend throughout their evolution was the freeing up of the shoulder girdle to have several ossified components, allowing greater movement in the pectoral fins. Perhaps such manoeuvrability was developed in order to chase the schools of plankton or ostracodes on which they fed, or it simply evolved as a more efficient way of escaping their attackers.

The acanthodiforms first appear in the Early Devonian and are represented by a single primitive genus, *Mesacanthus*, the only acanthodiform to have intermediate fin spines present between the pectoral and pelvic fins. *Mesacanthus* (meaning "middle spine") is known from well-preserved complete specimens from Angus, Scotland. By the Middle and Late Devonian the acanthodiforms were flourishing, and had achieved a standard, conservative body plan much like that of the

▲ Jaw bone of the ischnacanthiform *Taemasacanthus erroli*, from the Early Devonian limestones of Taemas, New South Wales, Australia. This specimen is about 4 cm long.

▼ Fin spines of *Gyracanthides* a large acanthodian from the Early Carboniferous of Mansfield, Victoria, Australia.

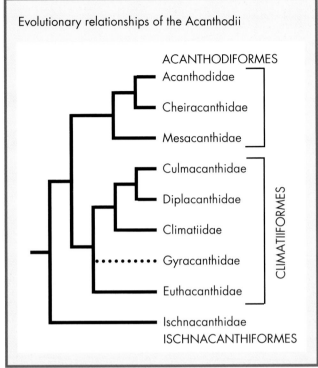

Evolutionary relationships of the Acanthodii

ACANTHODIFORMES
Acanthodidae
Cheiracanthidae
Mesacanthidae

Culmacanthidae
Diplacanthidae
Climatiidae
Gyracanthidae
Euthacanthidae

CLIMATIIFORMES

Ischnacanthidae
ISCHNACANTHIFORMES

differences in size of pelvic fins and numbers of ribs on the fin spines. In all advanced acanthodiforms, the scales have flat, unornamented crowns and rather

ischnacanthids. *Cheiracanthus* was a widespread Middle Devonian form, whose characteristic scales are recognised from sites of this age around the world, although the only complete specimens are known from the Old Red Sandstone of Scotland. Late Devonian forms such as *Howittacanthus*, from Mount Howitt, southeastern Australia, and *Triazeugacanthus*, from the Escuminac Formation of Quebec, Canada, vary only in minor ways from the later forms such as *Acanthodes*, by

shallow bases, and the fin spines tend to have only one large median rib. Advanced acanthodiforms such as *Acanthodes* and *Utahcanthus*, from North America, also had three kinds of otolith, or ear-stone, one for each chamber of the membranous labyrinth. The presence of well-formed otoliths may have given these fishes a greater degree of balance or a faster sense of direction when swimming rapidly.

Acanthodes was a cosmopolitan genus that lived

from the Early Carboniferous to the Middle Permian, and is one of the only acanthodians in which the braincase and gill arches are preserved as impressions in material of *Acanthodes bronni* from the Lebach site, Germany. The numerous gill rakers present on the gills, and the wide gape suggest that it was an efficient filter-feeder.

Acanthodes is the only known Permian acanthodian, representing the last survivor of this once flourishing class of fishes.

The acanthodians most likely died out because of the rapidly increasing numbers of ray-finned fishes and sharks at this time, which were expanding into a great

▲ *Howittacanthus kentoni*, a Late Devonian acanthodiform from the Mount Howitt site in southeastern Australia.

▼ Reconstruction of *Howittacanthus* with mouth wide open for filter-feeding.

▼▼ *Mesacanthus*, one of the earliest known acanthodiforms, from the Early Devonian Old Red Sandstone of Scotland.

range of habitats and feeding niches. Today we can utilise the minute scales of acanthodians from Silurian and Devonian deposits to date the age of sediments and make correlations with other faunas. Acanthodian fossils are becoming increasingly more important in this regard, and there is much research to be done to refine correlation schemes based on their scales.

POWRIE. 1891
92. — 275

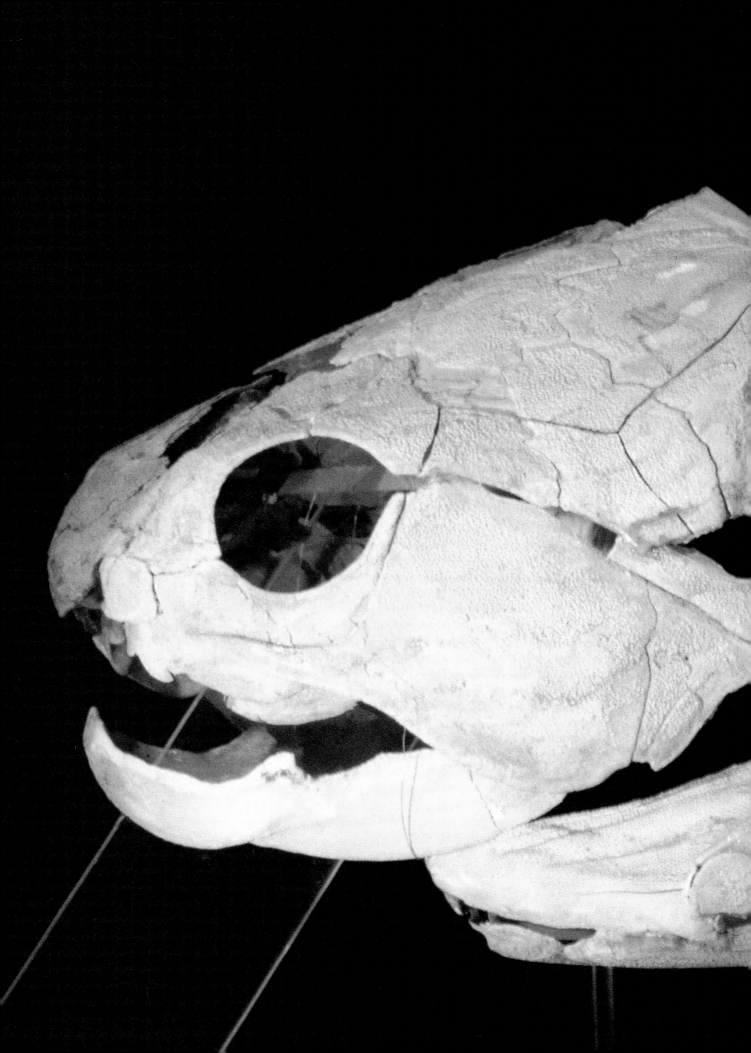

Chapter 5

ARMOURED FISHES AND FISHES WITH ARMS

CLASS PLACODERMI

The placoderms were armour-plated jawed fishes which first appeared early in the Silurian Period about 420 million years ago and dominated the seas, rivers and lakes of the Devonian Period, becoming extinct at the end of that age. They included some of the most bizarre vertebrates that ever lived, such as the antiarchs, characterised by external bone-covered arms, and the giant dinichthyids, the first creatures ever to reach gigantic sizes of 8 m or more. Placoderms sit on the evolutionary tree somewhere between the sharks and the true bony fishes. Their rapid evolution and diversification means that species generally occupied short timespans, and their fossilised bones can often be used to accurately determine the precise ages of Devonian rocks. The world's best-preserved placoderms undoubtedly come from Australia. Beautiful specimens of primitive placoderms from the Early Devonian limestones of southeastern Australia show extraordinary detail of the anatomy of these primitive jawed fishes; and a great diversity of advanced placoderms, three-dimensionally perfectly preserved from the Gogo Formation of northern Western Australia, show the complex adaptations evolved by these fishes for life in a reef ecosystem.

The word placoderm comes from the Greek, meaning "plated skin", alluding to their characteristic feature of having the head and trunk covered in a mosaic of overlapping bony plates. Placoderm fossils have been known for hundreds of years, most notably those from the Old Red Sandstone outcrops of Scotland, although some of the early reconstructions of placoderms depict them as jawless fishes with wing-like arms and unusually large heads. Hugh Miller, in particular, is famous for his "rude drawings", one being an 1838 composite restoration

Previous page: The armour-plated placoderm fishes ruled the seas, rivers and lakes for nearly 60 million years during the Devonian Period as the most successful group of fishes. This specimen is one of the superb, three-dimensionally preserved specimens of an arthrodire, *Eastmanosteus calliaspis*, from the Late Devonian Gogo Formation, Western Australia. These sites have yielded not only the world's best preserved placoderm fossils but also a highly diverse fauna containing about 25 placoderm species. *John A. Long*

Inset: The extraordinarily preserved placoderms from the Gogo Formation, in the far north of Western Australia, are found as nodules lying on the surface of the plains, between the fronts of the ancient reefs. The author is seen here searching for fish fossils. *R.E Barwick*

confusing two different Scottish placoderms, *Coccosteus* and *Pterichthyodes*. As the specimens were studied in more detail, the facts emerged that placoderms had true jaws with teeth and had fins much like other fishes. The earliest detailed study of placoderms was by the noted Swiss palaeontologist Louis Agassiz, who published a five-volume set entitled *Recherches sur les Poissons Fossiles* (Research on Fossil Fishes) between 1833 and 1843, which illustrated many forms of placoderms. Some early naturalists even regarded the antiarch placoderms as invertebrates akin to large beetles or as kinds of fossil turtles because of the box-like shell on their backs.

The revolution in placoderm studies came in the early 1930s when the Swedish Professor Erik Stensiö, of the Natural History Museum in Stockholm, began looking at placoderms in great anatomical detail. Stensiö took specimens of placoderm skulls with braincases intact in the rock and slowly ground them away, recording each cross-section of a tenth of a millimetre in wax templates, each magnified by ten times the original section size; when these were assembled they made a large model of the three-dimensional skull clearly showing canals for nerves, arteries and veins. This

method was called Sollas' grinding technique. By taking the negative of where bone was missing, Stensiö could then make a model of the brain cavity with soft tissues shown emerging from parts of the brain. His detailed works began a new era in placoderm work, and fossil fish studies in general, by describing their soft anatomy and placing placoderms as true jawed fishes closely related to modern-day sharks.

In recent years the acid-etching technique has enabled us to prepare specimens out of the rock in three-dimensional form, thus confirming much of Stensiö's work, and discovering much new information on placoderm anatomy and relationships.

Today placoderms are an important group for solving many geological problems. They are useful as index fossils for giving age determinations for Devonian sediments, and some groups have distinct biogeographic ranges, and therefore tell us about the positions of certain continents in past geological times. The debate over placoderm relationships has hotted up in recent years, with a number of new viewpoints being published. Some workers agree with Stensiö that placoderms should be classified with sharks and rays (in a superclass called the Elasmobranchiomorphii); others believe placoderms are closely related to the bony fishes or Osteichthyes; and yet others believe they occupy a position more primitive than either of these groups, at the base of the radiation of all jawed fishes.

▲ Hugh Miller's late nineteenth century "rude drawings" of Scottish placoderms. In those days the placoderms were seen as a curiosity, and were thought to be jawless fishes like the 'ostracoderms'.

◀ The famous Achanarras quarry, Scotland, where many fine placoderm fossils have been found over the past century from the Old Red Sandstone exposures.

Dr Nigel Trewin, Aberdeen University

GOGO FISHES AND THE ACID PREPARATION TECHNIQUE

The world's best placoderm fossils come from the Late Devonian reef deposits on Gogo and Christmas Creek stations in the Kimberley district of north Western Australia. The Gogo Formation represents the quiet deeper waters well away from high-energy reef fronts. The great diversity of fishes found in these deposits are shown throughout this book, preserved in exquisite detail because of the lack of later geological activity that kept the region free of large crustal movements which usually deform and compress fossils of this age.

In the deeper inter-reef basins of the ancient reef system, muddy lime-rich sediments slowly accumulated over the dead bodies of organisms, many of which may well have lived on or around the reef when alive. After death (by whatever means) fish carcasses tend to float for a while, then sink to the muddy sea floor, and may be scavenged upon at any time during this period. Thus the remains of Gogo fishes found in the fine mudstones comprise complete skeletons, isolated bones or pieces of a carcass. The Gogo fish were rapidly encased in fine limy mud, which set as hard calcite crystals formed not long after burial. This process protected the delicate skeletons from being crushed by the weight of accumulating overlying sediments.

Searching for Gogo fishes in the field is a matter of hard work, hitting thousands of limestone nodules with a hammer until a lucky find is made. Sometimes a keen eye can spot a bit of bone weathering out of a concretion, and in such cases the limestone nodule doesn't have to be broken. If a specimen of a complete fish has been split through the middle or broken by hammer into a handful of pieces, the specimen may be prepared using several techniques. One method involves embedding each side of the split fish in an epoxy resin slab and then acid-etching each half. This reveals the two sides of the fish with all bones in articulated position, as the fish was buried. Alternatively the pieces can be glued together with acid-resistant epoxy resin and then the whole nodule dissolved in acid. This gives a whole, undamaged skeleton, except for visible lines in some bones where the original breaks occurred.

The acid etching process uses weak acetic acid (about 10% solution) and because each treatment requires extensive washing in running water, the whole preparation procedure may take several months for each specimen. During the preparation, after each treatment in acid and washing in water, the exposed bones are left to dry in the air. They are then impregnated with a plastic glue that is absorbed into the bones and gives them internal strength. When the bones have soaked up as much plastic glue as necessary, the specimens can then be placed back in acid for further dissolution of the enclosing rock.

Once the fossil fish plates are freed from the rock, they can then be assembled (like building a model aeroplane) to restore the placoderm's external skeleton in perfect three-dimensional form. This chapter uses many examples of the superb Gogo placoderms to show their overall appearance and anatomical features.

John A. Long

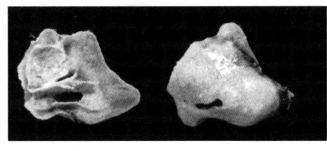

John A. Long

▲ Annular cartilages found intact on the snout of an arthrodire placoderm from Gogo, Western Australia. It is a further piece of evidence linking placoderms with chondrichthyans.

◄ Devonian reefs rise above the barren plains of Gogo Station in the far north of Western Australia.

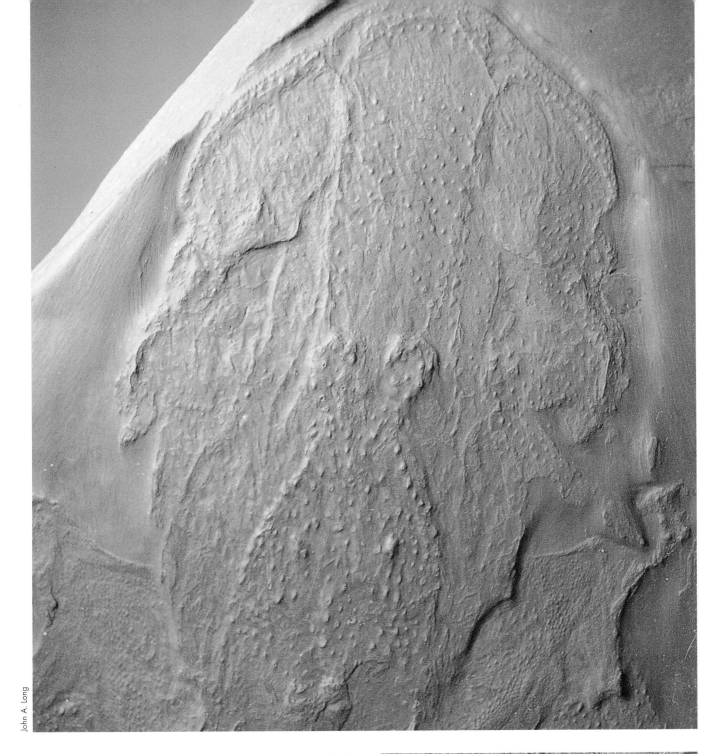

Image credit: John A. Long

▲ *Stensioella* from the Early Devonian Hunsrückscheifer (black shales) of Germany. This fish is regarded as the most primitive placoderm yet found as it lacks many of the plates present in the armour of other species.

▼▶ Male (right) and female (below) individuals of *Rhamphodopsis* from the Middle Devonian of Scotland. Note the male claspers visible just above the 6 and 7 numerals.

Image credit: John A. Long

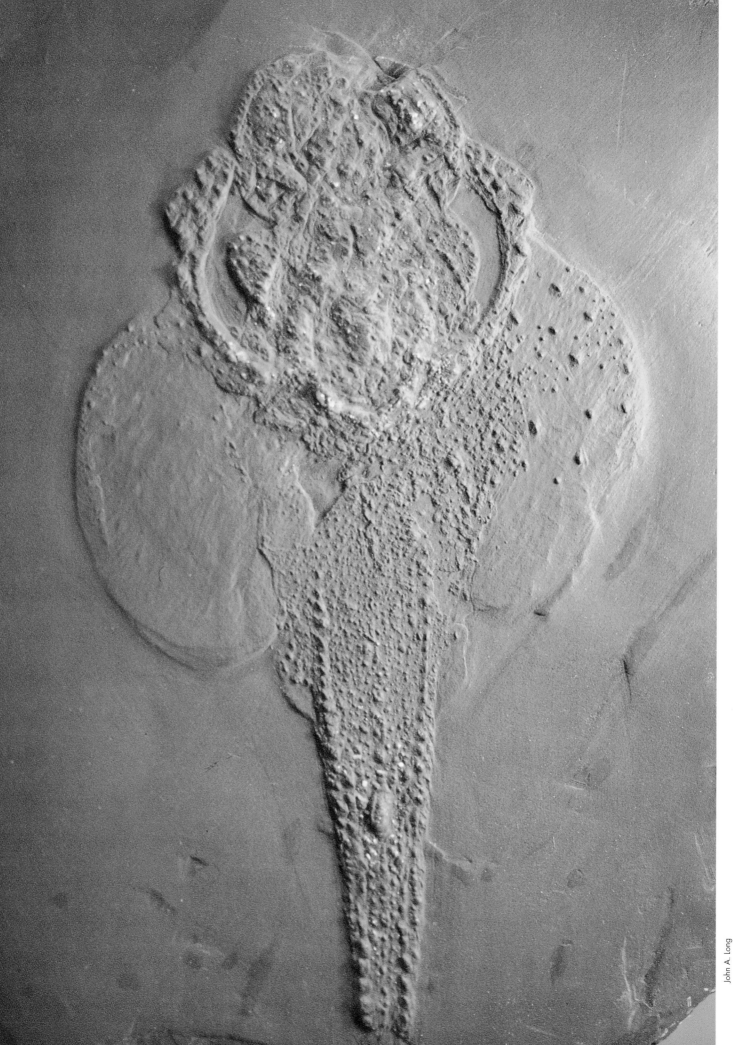

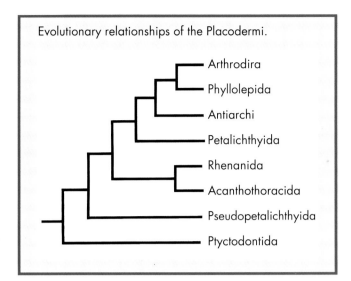

Evolutionary relationships of the Placodermi.

- Arthrodira
- Phyllolepida
- Antiarchi
- Petalichthyida
- Rhenanida
- Acanthothoracida
- Pseudopetalichthyida
- Ptyctodontida

PLACODERM ORIGINS AND RELATIONSHIPS

The placoderms arose about 420 million years ago, and by Early Devonian (about 400 mya) all the major placoderm orders had appeared. The affinities of the placoderms have long been a topic of debate among palaeontologists. Swedish scientist Erik Stensiö favoured a close relationship between placoderms and sharks, based on his many reconstructions of the anatomy of placoderms from the detailed serially-ground sections. However, many people involved in research on placoderms today have claimed that Stensiö based his anatomical reconstructions of placoderms on a shark model, thereby artificially making them always appear more shark-like. This led some scientists to suggest that the placoderms could be closely related to the ancestors of the true bony fishes (osteichthyans). Others have argued that placoderms evolved well before sharks, acanthodians or bony fishes and are the primitive ancestral group to all jawed fishes.

New evidence is now emerging that favours the original hypothesis of Stensiö that sharks and placoderms evolved from a common stock. One of the strongest pieces of original evidence uniting the two groups is that both have external clasping organs in males for internal fertilisation, although only one group of placoderms, the ptyctodontids, actually have these organs. It has been suggested that external claspers were a specialised feature, which evolved in sharks and all placoderms. This implies that claspers were primitively present in all placoderm groups but

◀ *Gemuendina*, a flattened ray-like rhenanid placoderm found in the Early Devonian black shales of the German Rhineland.

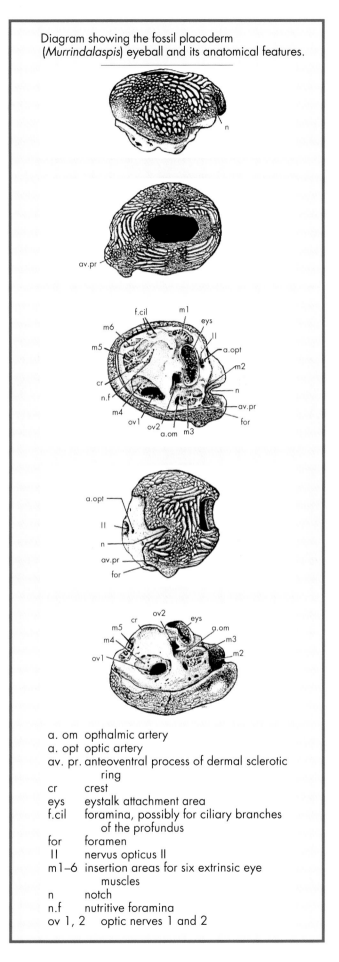

Diagram showing the fossil placoderm (*Murrindalaspis*) eyeball and its anatomical features.

a. om   opthalmic artery
a. opt   optic artery
av. pr. anteoventral process of dermal sclerotic ring
cr   crest
eys   eystalk attachment area
f.cil   foramina, possibly for ciliary branches of the profundus
for   foramen
ll   nervus opticus ll
m1–6   insertion areas for six extrinsic eye muscles
n   notch
n.f   nutritive foramina
ov 1, 2   optic nerves 1 and 2

THE BASIC STRUCTURE OF PLACODERMS

Placoderms are characterised by their armour made of overlapping bony plates forming a head shield and a trunk shield. These mostly articulate by a knob and groove or may be fused as one composite shield. Each of the seven groups of placoderms has a unique pattern of bony plates, and each plate has a specialised shape, overlap regions, surface texture (ornament) and possibly sensory-line grooves. Primitive placoderms have wart-like ornament (tubercles), advanced forms may have linear ridges and network patterns.

The braincase of all primitive placoderms was well ossified with layers of laminar perichondral bone, but in later species it may be formed entirely of cartilage, an adaptation to possibly reduce weight. The jaws are always simple rods of bone, which may have pointed tooth-like structures for gripping prey or, in those species whose diets involve crushing hard-shelled prey, the jaws may have areas of thickened tubercles. The jaw joint is simple; the lower jaws articulate against a knob-like quadrate bone that is commonly fused to the cheek bones. The eyes are surrounded by a simple ring of bone, made up of three to five parts called sclerotic plates, and are connected to the braincase by an eyestalk. The head-shield itself may take a variety of bone patterns, some with eye notches to the side of the head, others

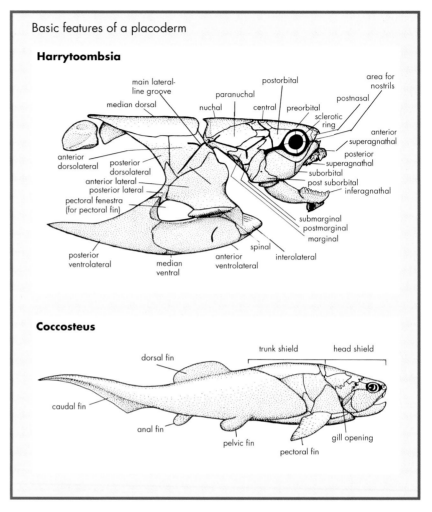

Basic features of a placoderm

Harrytoombsia

Coccosteus

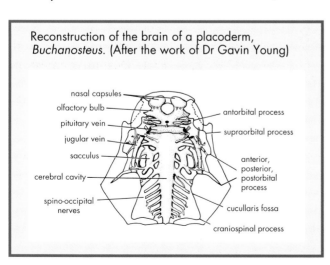

Reconstruction of the brain of a placoderm, *Buchanosteus*. (After the work of Dr Gavin Young)

having eyes and nostrils together in the centre of the skull.

The body of placoderms is generally torpedo-like, as in sharks, although notable exceptions are seen in the flattened groups, the phyllolepids and the rhenanids. There is only one dorsal fin developed—despite many erroneous reconstructions made by Erik Stensiö, in which two dorsal fins were always assumed to be present. There are paired dorsal and pelvic fins with shark-like shape and internal structure and, in many species, a single anal fin. The body is primitively covered with thick bony platelets which resemble miniature versions of the dermal bones, often having similar ornament on each scale. In advanced lineages of placoderms the body scales may be reduced or absent, corresponding with the overall trend seen throughout the skeleton to reduce weight. Placoderms became extinct at the end of the Devonian, some 355 million years ago, after dominating life in the seas, rivers and lakes for some 60 million years—one of the most successful groups of fishes ever known.

were independently lost in later placoderms except the ptyctodontids.

Other anatomical features that suggest close affinity between sharks and placoderms are the presence of an eyestalk connecting the eyeball to the braincase; the structure of the pectoral and pelvic fins, which are fleshy broad-based structures (except in antiarchs, in which they are bony props); and the similar shapes of the braincases in primitive sharks and placoderms. The presence of little cartilage rings around the nasal area, called annular cartilages, was proposed as an important similarity between sharks and placoderms by Stensiö. Although he never saw such a structure in any fossil, he based his theory on similarities in the general appearance of the snout region in the two groups. Quite recently a superbly preserved placoderm from Gogo, Western Australia, has shown that annular cartilages really did exist in some placoderms. The general shape of the bodies in predatory placoderms, such as *Coccosteus,* is reminiscent of sharks and suggests a similar swimming style and internal anatomy (the absence of a swim-bladder) in both groups.

There are some 200 known genera of placoderms divided into seven major orders, the largest order of placoderms being the Arthrodira ("jointed necks"), comprising more than 60 per cent of all known placoderm species. All these groups had appeared by the start of the Devonian and most survived through to the end of that period. The different orders are recognised by a combination of plate patterns making up the external armour, and by the unusual and often complex development of bony surficial ornament on the plates of some groups.

A view widely held by placoderm specialists is that the ptyctodontids, rhenanids and acanthothoracids are the most primitive placoderms, whereas the antiarchs, arthrodires and phyllolepids are considered the most advanced orders. The others slot somewhere in between these, although as several schemes of interrelationships have been proposed in recent years, there is still ongoing debate about the precise relationships of certain orders, especially the antiarchs and petalichthyids.

STENSIOELLIDA AND PSEUDOPETALICHTHYIDA: EARLY ENIGMATICS

The Stensioellida and Pseudopetalichthyida are primitive orders of placoderms with little bony armour developed. Their broad wing-like pectoral fins are reminiscent of rays, and their bodies have many small ornamented denticles set in the skin. Relationships of these groups is still not clear, although they are regarded as the most primitive of all placoderms as they lack a number of bones found in all other groups. The group is only known from a few species, and all known specimens are from the Early Devonian black shales (Hunsrückschiefer) of the German Rhineland.

RHENANIDA AND ACANTHOTHORACI: THE ELABORATELY ARMOURED ONES

The Rhenanida are a group of flattened placoderms similar to stensioellids in having very large wing-like pectoral fins. The skull is often formed of many small bones and may not resemble the larger mosaic patterns of other placoderms. The trunk shield is very short, and the tail has many bony platelets of varying sizes. The group is known only from Europe and North American marine sediments of Early Devonian age. Rhenanids are now often grouped with the Acanthothoraci (meaning "spiny trunk shield"), a group characterised by heavily ossified armour with elaborate ornamentation. Indeed, the fossil plates of some of these fishes have the most beautiful surface patterns ever seen in the dermal skeleton of any vertebrate. Acanthothoracids are characterised by their own patterns of skull bones and short trunk shield.

Australia has an excellent representation of fossil acanthothoracids, from the Early Devonian limestones around Taemas and Wee Jasper, in New South Wales, and near Buchan, in Victoria. These are three-dimensionally preserved skulls and trunk plates, and often include surprising preservation of structures like the

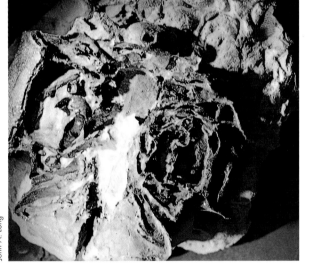

John A. Long

▲ A skull of the placoderm *Buchanosteus,* which has been weathered just enough to expose the delicate tubes of the braincase, as preserved in thin layers of perichondral bone (see diagram, facing page).

▲ The skull of *Brindabellaspis stensioi*, an Early Devonian acanthothoracid placoderm from the Taemas–Wee Jasper region of southeastern Australia. This recently found specimen shows the long snout previously undetected from earlier finds.

▶ The skull of the petalichthyid *Shearsbyaspis* seen in visceral view. This placoderm comes from the Early Devonian of southeastern Australia.

▶ Trunk plates of the Late Devonian ptyctodontid *Campbellodus decipiens* from Gogo, Western Australia. *John A. Long*

▼ A Middle–Late Devonian ptyctodontid placoderm, *Ctenurella gladbachensis*, from the Bergisch–Gladbach area of Germany.

bony covering around the placoderm eyeball. One such fossil, a sclerotic capsule, found near Taemas, demonstrated that the eyes of placoderms had the same pattern of complex eye musculature and nervous, arterial and venous anatomy as do modern fishes. Unlike modern fishes, however, the eyes were heavily invested with bone, and each eyeball was connected to the braincase by a pedicle of bone or cartilage called the eyestalk. The eyestalk is another feature seen in both sharks and placoderms.

The best-known acanthothoracid fossils from southeastern Australia are the remains of the long-snouted *Brindabellaspis* and the two high-crested forms *Weejasperaspis* and *Murrindalaspis*. These have exquisite preservation of the cavities surrounding the brain and cranial nerves, and reveal much about the soft anatomy of primitive placoderms. Other well-preserved acanthothoracid fossils occur in the Early Devonian sediments of eastern Europe, Russia and Arctic Canada.

PTYCTODONTIDA: CRUSHING TEETH AND SPIKY CLASPERS

The Ptyctodontida (meaning "beaked tooth") were an unusual group of placoderms with strong crushing plates in the jaws. They had long bodies with whip-like tails, and large heads with big eyes and in many respects resembled the modern day chimaerids and whipfishes. They had very short trunk shields and reduced head bone cover, and were specially adapted for feeding along the bottom of the seas on hard-

shelled organisms. They are the only placoderms to show sexual dimorphism—the males having dermal clasping organs.

Ptyctodontid fossils in Australia are known from isolated tooth-plates of Early Devonian age from the limestones near Taemas—Wee Jasper, in New South Wales and from extremely well-preserved articulated skeletons of Late Devonian age from the reef deposits at Gogo, Western Australia. There are two forms from Gogo. *Ctenurella gardineri* was about 20 cm long and had

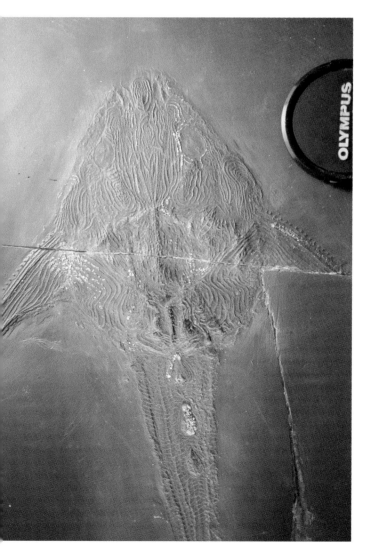

▶ Reconstruction of *Lunaspis* (in dorsal view), one of the early petalichthyid placoderms whose remains have been found in Germany, China and Australia.
John A. Long

▼ *Lunaspis*, one of the best known of the petalichthyid placoderms. This nearly complete specimen comes from the Early Devonian Hunsrückschiefer of the Rhineland, Germany.

delicate thin scales covering the body, and a relatively low trunk shield. *Campbellodus decipiens* was about 30–40 cm maximum length and had an unusual trunk shield with three plates forming a high spine on its back, preceding the dorsal fin. Specimens of *Ctenurella* from similar-age rocks in Germany have whole bodies preserved which reveal the outline of the fish and the position of the fins, and these have been used to give the new restorations shown in this book. It is interesting that the Gogo ptyctodontids show the body covered by fine overlapping scales, whereas other specimens from around the world lack this body scale cover. The Gogo specimens also include parts of the braincase, indicating that it was highly modified for placoderms—not a single bony box with holes for nerves and arteries, but a complex unit made up of several ossifications.

Ptyctodontid fossils are also well-known from Middle and Late Devonian deposits in Europe, North America and Russia. Whole skeletons of the little ptyctodont *Rhamphodopsis* have been described from Scotland and were the first studied examples showing the sexual dimorphism in placoderms by English palaeontologists D.M.S. Watson and Roger Miles. In many cases ptyctodontids are often represented solely by their characteristic tooth-plates. Such isolated fossils from North America indicate that the largest species of ptyctodontids (*Eczematolepis*, Late Devonian of New York) had tooth-plates about 15–20 cm long , suggesting a total fish length of about 2.5 m.

PETALICHTHYIDA AND THEIR ODD RELATIVES

The Petalichthyida were unusual little placoderms with widely splayed pectoral fins and all their dermal bones ornamented with characteristic linear rows of little tubercles. The bones have thick tubes carrying the sensory-line nerves, and these tubes clearly stand out on the inside surface of the skull bones. Petalichthyids are known from Europe, North and South America, Asia and Australia. They reached their peak of diversity in the Early Devonian and only a few species survived until the Late Devonian. Unusual petalichthyid-like fishes called "quasipetalichthyids" are known from China, which indicate a minor local radiation of the group took place on the isolated South China continental block.

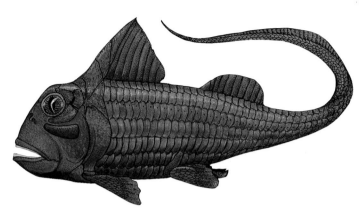

▲ Reconstruction of *Campbellodus decipiens*, after a new complete specimen from the Gogo Formation, Western Australia, which shows large scales covering the body.

Some of the best preserved skulls of petalichthyids come from the Early Devonian limestones near Taemas, New South Wales, Australia. Several forms have been described from this region, including *Notopetalichthys*, *Shearsbyaspis* and *Wijdeaspis*, the latter also known from Europe and Russia. Probably the best fossils of petalichthyids, preserved as whole fishes, come from the Early Devonian black shales of Germany. *Lunaspis* is the best-known example, and its remains have also been identified from China and Australia. Petalichthyids were probably bottom-dwelling fishes that swam slowly searching for prey on the sea floor. Unfortunately none of the known petalichthyids have mouth parts preserved, so we can only guess at their eating habits.

THE FLAT AND FEARSOME PHYLLOLEPIDA

The Phyllolepids (meaning "leaf scale") were flattened, armoured forms, long thought to be jawless fishes, until Professor Erik Stensiö demonstrated in 1936 that they were true placoderms. In 1984 I described whole complete phyllolepid specimens from Mount Howitt, southeastern Australia. These actually had the toothed jaw bones preserved, leaving no doubt as to their placoderm affinity. Phyllolepids are readily distinguished by their flat armour made up of a single large plate on the top of the head and trunk rimmed by a regular series of smaller plates. Their most characteristic feature is that each plate has a radiating pattern of raised fine ridges and tubercles.

Only one genus, *Phyllolepis*, was known until recent discoveries from Australia and Antarctica began filling in the gaps of phyllolepid evolution. *Phyllolepis* is

known from the Late Devonian (Famennian stage) of Europe and North America, whereas the Australian phyllolepids, *Austrophyllolepis* (meaning "southern leaf scale") and *Placolepis* (meaning "plate scale") appeared earlier in time (Givetian—Frasnian), and appear to be far more primitive than *Phyllolepis*. Another new genus of phyllolepid was discovered in 1993 near Canowindra, New South Wales, Australia, by Dr Alex Ritchie of the Australian Museum. Like *Austrophyllolepis* it has a median ventral plate but differs in its head shield plate arrangement.

Austrophyllolepis is known from two species from Mount Howitt, eastern Victoria, and its fragmentary remains have also been found in Antarctica. These specimens have revealed more about the anatomy of

▲ The skull of *Quasipetalichthys*, a member of an unusual group of petalichthyid-like placoderms found only in China. The photo is of a cast of the skull.

▲ An isolated *Phyllolepis* plate from the Late Devonian of East Greenland. The striking surface pattern of concentric ridges makes phyllolepid plates easy to recognise.

▶ The complete armour seen in dorsal view of a Late Devonian phyllolepid placoderm, *Phyllolepis woodwardi*, from the Dura Den locality in Scotland.

▶ Reconstruction of *Austrophyllolepis*, a phyllolepid placoderm known from well-preserved whole specimens from Mount Howitt, Victoria, Australia.

▼ A Late Devonian phyllolepid, *Austrophyllolepis ritchiei*, from Mount Howitt, Victoria. Remains of these fishes from Mount Howitt show details of the tail, jaws, parasphenoid and otoliths. This specimen is a whitened latex peel taken from the cleaned impression of the fish remains preserved in shale.

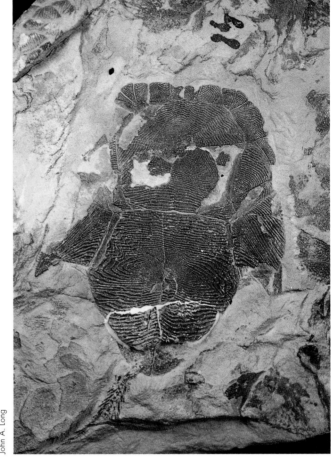

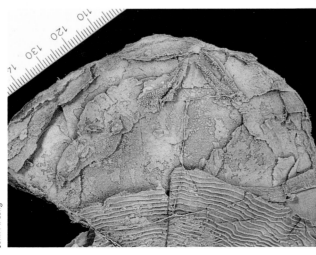

phyllolepids than any other site, as they show the outline of the whole fish, along with preserved impressions of the jaws, pelvic girdles, palate bone (parasphenoid), and parts of the cheek. From study of these it is possible to build a picture of phyllolepids as flat predators, lying in wait on a muddy lake bed waiting for some unsuspecting fish to swim above. Then, using its unusually long tail, it could thrust upwards to catch the prey in its set of gripping jaws. Phyllolepids were probably blind, as they lacked bones circling the eyes and the head shield is not embayed for the eyes as in most placoderms. They had especially well-developed radiating patterns of sensory-line canals which may have helped them detect prey swimming above them while they were concealed below the surface under a fine layer of sediment. The largest phyllolepids were only about 50–60 cm in length. The study of the new specimens from Australia suggests that phyllolepids might even be primitive arthrodires, similar to unusual forms like *Wuttagoonaspis*.

ARTHRODIRA: THE GREAT PLACODERM SUCCESS STORY

The Arthrodira (sometimes called Euarthrodira) are the only group of placoderms to have two pairs of upper jaw tooth-plates (called superognathals). The skull has a regular pattern of bones, featuring eyes located to the sides of the head, and a separate cheek unit that hinged along the side of the skull roof. The head and trunk shields are joined by a ball-and-socket type joint in all advanced arthrodires; as in primitive groups like actinolepidoids, they have a sliding neck joint. The arthrodires had shark-like bodies with a single dorsal fin, broad fleshy paired pectoral and pelvic fins, and an anal fin. The tail was primitively covered in scales, but naked in many advanced forms. There were many different families of arthrodires but in general they are divided into the primitive groups having long trunk shields with large spinal plates, and the advanced groups having shortened trunk shield with reduced spinals, and some with pectoral fins not fully enclosed by the trunk shield without a spinal plate. Arthrodires are known throughout the Devonian rocks of the world.

The most primitive arthrodires were the long-shielded actinolepids, such as *Dicksonosteus* from Spitzbergen, which flourished in the Early Devonian shallow seas and rivers, especially in Euramerica.

These arthrodires lacked a true ball-and-socket neck joint, instead having a simple sliding neck joint where the head shield sat on a flat platform of bone protruding from the trunk shield. The phlyctaeniid arthrodires were the first arthrodires to evolve the primitive ball-and-socket joint, enabling greater vertical mobility of the head shield, and consequently better ability to open the mouth and catch prey. Some of these, such as the Early Devonian *Lehmanosteus* and *Heintzosteus*, described from Spitzbergen by Dr Daniel Goujet, have very widely splayed spinal plates and look very streamlined. Their jaws had many small teeth,

▶ A large incomplete arthrodire skull from the Early Devonian limestones exposed at Burrinjuck Dam, near Wee Jasper, New South Wales, Australia. Preserved length is 27 cm.

▼ An Early Devonian marine scene showing the fishes and invertebrate life of the Taemas–Wee Jasper fauna, New South Wales. The placoderm in the centre is *Buchanosteus*, a primitive arthrodire about 30 cm long. It is chasing a small *Nostolepis* acanthodian. Below these fish to the left is a small onychodontid crossopterygian, and below this fish is the unusual high-crested acanthothoracid placoderm *Murrindalaspis*. The large fish in the bottom right corner is the lungfish *Dipnorhynchus*, and the small fish to the top right is one of the first ray-finned fishes, *Ligulalepis*. Typical invertebrates of the time include the straight orthoconic nautiloid (centre left), trilobites (to the rear, above the nautiloid head), horn-shaped rugose corals and large dome-like tabulate corals, and crinoids or sea-lilies (far right and far left).

John A. Long

like phyllolepids, and were probably used for gripping soft-bodied worms and such.

Australia has an excellent fossil record of arthrodires beginning with superb three-dimensional plates and skulls of Early Devonian age from near Taemas in New South Wales and Buchan, Victoria. Numerous species have been discovered in these rocks, ranging from small forms about 20 cm long to giants with skulls nearly 40 cm long, suggesting a whole body length of up to 3 m.

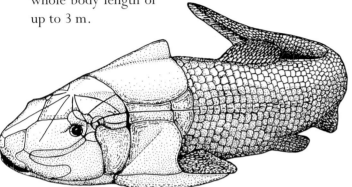

The most common species from southeastern Australia is *Buchanosteus confertituberculatus*, named after Buchan, and the species name meaning "rounded tubercles". The species was first described by Frederick Chapman of the National Museum of Victoria, who misidentified it. Professor Edwin Hills, of the University of Melbourne, recognised it as an arthrodire and pioneered a method of preparing out the specimen using acid. Unfortunately he used hydrochloric acid, which also damaged the specimen, but not before he extracted a lot of new information on the structure of the braincase. Although Hills recognised the species as an arthrodire, he cautiously assigned it to the well-known Scottish genus *Coccosteus*. Erik Stensiö, of Sweden, in 1945 reassigned Hill's specimen to a new genus, which he called *Buchanosteus*. In the late 1970s Dr Gavin Young, from the Australian Geological Survey Organisation, published a detailed study of the anatomy of *Buchanosteus*, reconstructing much of the braincase and cranial anatomy from new specimens prepared using acetic acid. Using this data he was able to propose a new theory of arthrodire interrelationships and thus heralded a new era for the study of placoderms. Buchanosteid arthrodires have now been found in China and Russia, and they are regarded as amongst the most primitive family of the great radiation of "advanced" arthrodires having a shortened trunk shield.

John A. Long

▲ Impressions of the bony plates of the unusual arthrodire *Wuttagoonaspis*, preserved in sandstones of Early–Middle Devonian age, found near Cobar, New South Wales, Australia.

◄ Reconstruction of *Wuttagoonaspis*, a primitive arthrodire known from several sites in central and southeastern Australia.

▼ Braincase of *Wuttagoonaspis*, preserved as an internal mould from the visceral surface of the skull.

John A. Long

Other placoderms found with *Buchanosteus* include the elaborately ornamented *Errolosteus*, named in honour of Errol White, a palaeontologist at the British Museum of Natural History who pioneered studies of

REPRODUCTION AND THE ORIGINS OF VERTEBRATE SEX

Reproduction in fishes generally involves the male's shedding sperm over the female's eggs in the external environment, although internal fertilisation is the norm for chondrichthyans and some placoderms and has independently evolved in some higher actinopterygians (certain teleosts). In external fertilisation the eggs and sperm are shed through the cloacal opening or urogenital sinus, although there are various different methods for the ova to be moved from the ovaries to the outside of the female's body. Internal fertilisation in chondrichthyans takes place by the male intromittent organs (claspers, modified structures at the base of the pelvic fins), which transmit sperm directly inside the female. Usually fishes using internal fertilisation have well organised mating behaviour and copulatory rituals. Many of the modern higher actinopterygians also have sophisticated mating behaviour and are territorial, building nests before external fertilisation takes place.

Some very primitive fishes, such as the hagfish, are hermaphroditic — the front half of the gonad is ovarian, producing eggs, and the rear half is testicular, producing sperm. During growth, one or other half eventually predominates, although in some cases individuals can remain sterile throughout life. In most fishes the ovaries are paired, although one side may become more developed than the other, or one side atrophies, as occurs in many chondrichthyans.

Sharks can lay eggs in two ways. Oviparous forms lay few large eggs, one at a time, which take a long time to hatch. Others are ovoviviparous, laying thin-shelled eggs inside the body which break down before birth, so that the shark appears to give birth to "live young" as in higher vertebrates. The gestation period of some sharks, such as *Squalus*, may be up to two years.

None of the early jawless fishes appear to exhibit visible sexual differences (termed sexual dimorphism where males and females have clear morphological differences). The first record of sexual distinction appearing in any fish is in the Middle Devonian ptyctodontid placoderm *Rhamphodopsis*, which had external clasping organs on males and wide pelvic basal plates on females, indicative of internal fertilisation and thus also of sexual intercourse in vertebrates taking place. Interestingly, ptyctodontids appear to be the only vertebrates whose males are armed with external bone covering the intromittent organ. The clasper has an external bony support rod along its length as well as a roughened tip with numerous small spiky projections. Such structures no doubt assisted the male during copulation and may have excited the female.

The fossil record of ptyctodontids actually dates further back to the earliest Devonian, and sharks, also known to have internal fertilisation, date further back into the Silurian Period, based on findings of fossilised scales. From these records it is possible to associate the earliest, and most primitive, jawed fishes having internal fertilisation, a feature not seen in any of the agnathans. Reproduction by external fertilisation in most fishes, such as acanthodians and osteichthyans, is generally regarded as being the primitive condition (also seen in agnathans), thereby implying that internal fertilisation is a specialised condition among some placoderms and chondrichthyans. However, as the males of most placoderms do not have claspers, this condition is thought to be secondarily lost in placoderms as longer trunk armours evolved, presumably making copulation increasingly difficult.

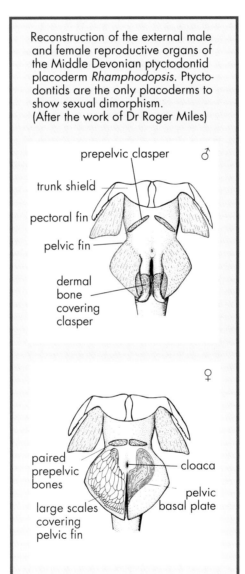

Reconstruction of the external male and female reproductive organs of the Middle Devonian ptyctodontid placoderm *Rhamphodopsis*. Ptyctodontids are the only placoderms to show sexual dimorphism.
(After the work of Dr Roger Miles)

prepelvic clasper ♂
trunk shield
pectoral fin
pelvic fin
dermal bone covering clasper

♀
paired prepelvic bones
large scales covering pelvic fin
cloaca
pelvic basal plate

Taemas–Wee Jasper placoderms; and *Arenipiscis*, a form with a slender skull roof and delicate sandgrain-like ornamentation (hence the name meaning "sand fish"). The largest arthrodire skull found at Taemas belongs to an as yet undescribed form. It has a large, flat skull about 40 cm long, with a small T-shaped rostral plate at the front—a feature linking the generally primitive arthrodires of the Early Devonian with the more advanced forms common in the Middle Devonian. The giant from Taemas may have been related to similar large forms known from Europe, called homosteids. These arthrodires had weak, toothless jaws and may have been early filter-feeders (like whale sharks and baleen whales).

Middle Devonian fish faunas of inland New South Wales and central Queensland contain the unusual arthrodire *Wuttagoonaspis*, named after Wuttagoona station near Cobar, New South Wales, where the fossil was first found. *Wuttagoonaspis* has a long skull with small distinct eye holes and a bizarre pattern of skull roof plates, most unlike any other arthrodire. When first described by Dr Alex Ritchie in 1973, the exact affinities of this fish were uncertain, and many workers still regard it as being in a group of its own, well apart from all the regular arthrodires. Giant *Wuttagoonaspis* plates found near the Toomba Range in southwestern Queensland, indicate that some species grew up to about 1 m long. The bones have a highly distinctive

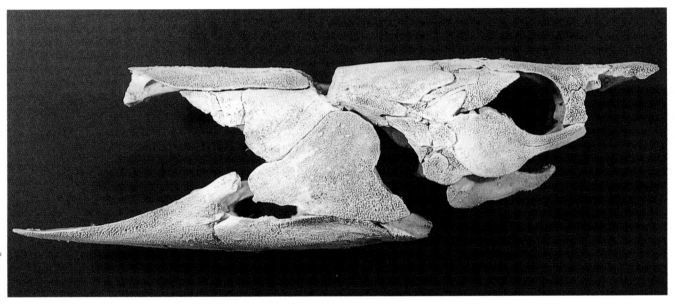

John A. Long

▲ *Fallacosteus turnerae*, one of the streamlined camuropiscid arthrodires found only at Gogo, Western Australia. The camuropiscids were probably fast-swimming top-water predators that chased the abundant shrimp-like crustaceans that lived around the ancient reefs.

◀ Skull and partial trunk shield of *Holonema westolli*, a long-shielded arthrodire from Gogo, Western Australia.

▼ View of the armour of *Holonema* seen from the side. Note the barrel-like shape of the body.

John A. Long

ornament, somewhat akin to the radiating pattern of ridges and tubercles also found in phyllolepids. The weak jaws of *Wuttagoonaspis* suggest that it may have been a bottom-feeding form, scrounging around in shallow estuarine or marine waters in search of small worms or other prey items.

The Late Devonian Gogo fauna of Western Australia includes more than 20 different arthrodires, some belonging in their own unique families. The camuropiscid and incisoscutid arthrodires are small forms that flourished on the ancient Gogo reefs; they are characterised by their elongate, spindle-shaped armours, large eyes and crushing tooth-plates. Extreme elongation of the armour is seen in forms like *Rolfosteus* and *Tubonasus*, which evolved tubular snouts to improve their streamlining. These fishes were only about 30 cm long, and may have been active top-water predators, probably chasing the small shrimp-like crustaceans that abounded in the

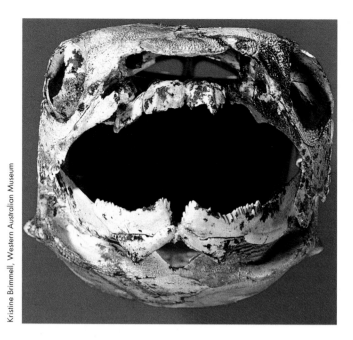

Kristine Brimmell, Western Australian Museum

▲ Face-on view of a new type of predatory arthrodire from Gogo, showing the jaws and three-dimensional shape of the armour.

▶ *Mcnamaraspis*, new predatory plourdosteid arthrodire from Gogo, Western Australia. This voracious little fish was about 25 cm long, and characterised by its very short spinal plates in front of the pectoral fins.

▼ *Holonema* lived around the reef at Gogo, and may have been coloured similar to modern reef fishes. This reconstruction is based on complete specimens found at Gogo, Western Australia.

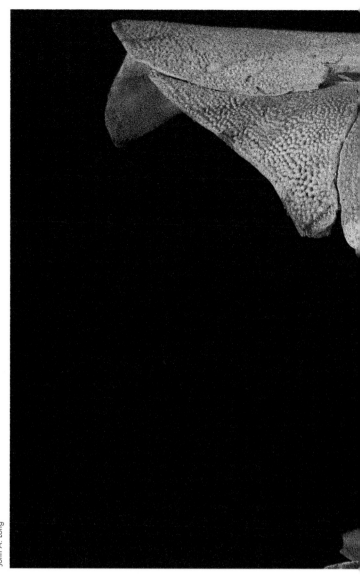

John A. Long

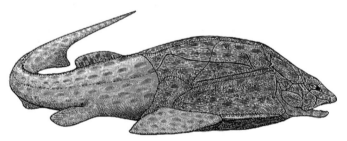

warm tropical waters. *Incisoscutum* had the trunk shield incised to free the pectoral fin from being enclosed by bone, probably improving the mobility of its pectoral fins. It was one of the commonest placoderms found at Gogo and also had strong crushing tooth-plates, like the camuropiscids.

The most diverse group of arthrodires found at Gogo are the little predatory plourdosteids. The group is named after *Plourdosteus*, a form that is well-known from the Late Devonian of Canada and Russia. Many different species of plourdosteid occur at Gogo, each distinguished by a unique pattern of skull roof bones, or dentition (for example *Torosteus*, *Harrytoombsia* and *Kimberleyichthys*). The lower jaws feature several well-developed cusps and toothed areas, with numerous

teeth along the midline where the jaws meet. In addition they have well-developed bony buttresses that braced the cartilaginous braincase for their powerful biting actions. They were obviously carnivorous little fishes, ranging in size from about 30 to 50 cm. The broad, robust shape of their armour suggests that they hunted near the sea floor or within the cavities of the ancient reefs, lunging out after small fishes or crustaceans.

The Gogo fauna includes large predators with huge dagger-like cusps on their jaws—such as *Eastmanosteus calliaspis*, which grew to about 3 m long, as well as many small predatory forms armed with sharp teeth-like structures on their jaws. *Eastmanosteus* shows us the basic pattern adopted by the largest placoderms—the dinichthyids (meaning "terrible fish"). Large skulls and bony plates of these monsters have been excavated from the Cleveland Shales of New York, and from limestones exposed in

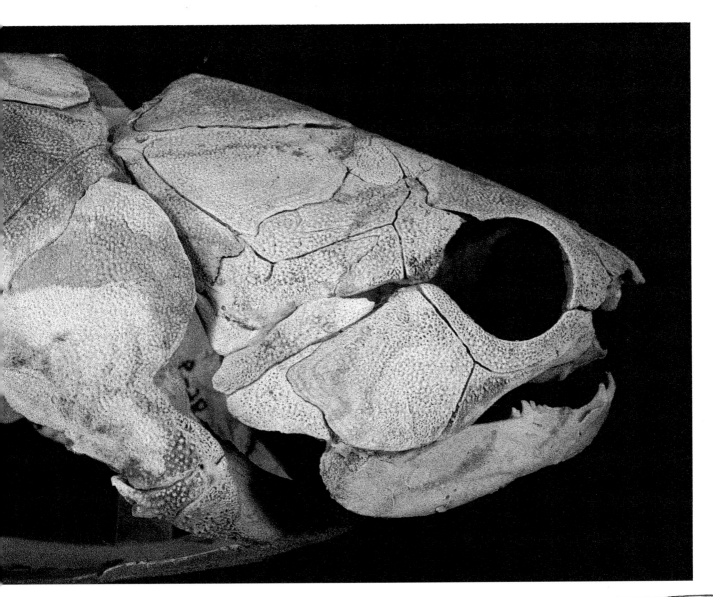

the northern Sahara Desert of Morocco. The largest of these had skull-roofs more than a metre long, which suggest total lengths of 6-8m. Most of these, like *Dunkleosteus* and *Gorgonichthys* had trenchant, pointed cusps on the lower and upper jaw tooth-plates. *Titanichthys*, from Morocco, was at least 7 m long but had weak jaws that lacked the sharp cusps of the dinichthyids, and it may have

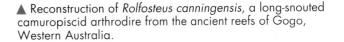

▲ Reconstruction of *Rolfosteus canningensis*, a long-snouted camuropiscid arthrodire from the ancient reefs of Gogo, Western Australia.

been a giant filter feeder like the modern-day whale shark. This type of feeding habit has also been suggested for an earlier group of large placoderms, the homosteids, by Dr Elga Mark-Kurik. Studies of the giant placoderms from the Cleveland Shale suggest that they may have preyed on other arthrodires or been carrion feeders. Perhaps they tried to catch the cladoselachian sharks that lived with them, but it seems hard to imagine the heavily armoured predators ever catching a streamlined little shark.

▲ Reconstruction of the late Devonian arthrodire *Groenlandaspis* from Mount Howitt, Victoria, Australia. This placoderm was first found in East Greenland (hence the name) in the late 1920s, and in the 1970s has been discovered in Antarctica, Australia, North America, Europe and Russia, and in 1994, from South Africa.

▶ A reconstruction of *Gorgonichthys*, a 6 m predatory placoderm from North America.

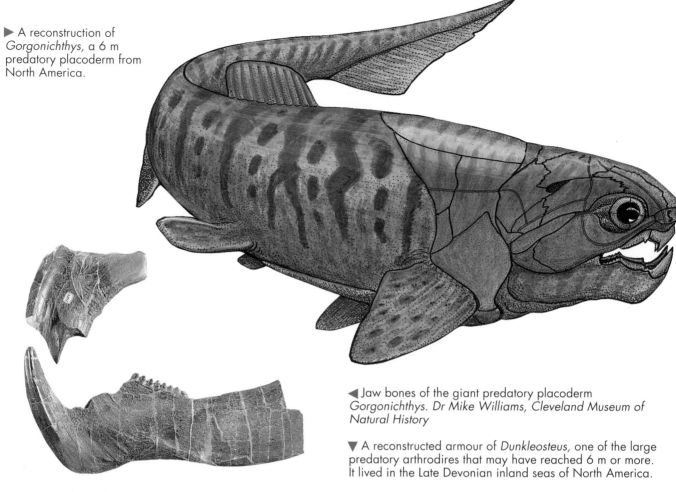

◀ Jaw bones of the giant predatory placoderm *Gorgonichthys*. *Dr Mike Williams, Cleveland Museum of Natural History*

▼ A reconstructed armour of *Dunkleosteus,* one of the large predatory arthrodires that may have reached 6 m or more. It lived in the Late Devonian inland seas of North America.

▲ Head shield, seen from above, of the antiarch *Asterolepis*, from the Late Devonian Lode site in Latvia.

▶ Skull and other remains of the primitive Early Devonian antiarch, *Yunnanolepis*, known from China and northern Vietnam.

▼ A pectoral appendage of a large species of *Remiogolepis* found near Eden, New South Wales, Australia.

THE ANTIARCHI:
ALWAYS AT ARM'S LENGTH

The antiarchs were little placoderms mostly about 20–30 cm in length, and reaching a maximum size of about 1 m. They are characterised by having the pectoral fins enclosed in bony tubes (often referred to as pectoral appendages). Most antiarchs had segmented arms, although one (*Remigolepis*) had short oar-like props. The head shield of antiarchs featured a single opening in the middle for the eyes and nostrils and pineal eye, and the trunk shield of all antiarchs was very long relative to the overall length of the fish. Antiarchs are unique in having a trunk shield with two plates along the back (median dorsal plates). The group first appeared in Late Silurian (found in China), and was widespread by the Middle Devonian, reaching a peak of species diversity during the Late Devonian.

The earliest and most primitive antiarchs are well represented in Lower Devonian rocks of Yunnan, China, from where they take their name, the yunnanolepidoids. These have a single prop-like pectoral fin, which has not developed the ball-and-socket articulation seen in all subsequent antiarchs. The armour of these little fishes, usually less than 5 cm long, has a covering of wart-like tubercles. Some more advanced forms have an incipient pectoral joint (in *Procondylepis* for example), and the true antiarch shoulder joint evolved from this basic structure.

The sinolepidoids (meaning "Chinese scale" forms) were a peculiar group of big-headed antiarchs with long, segmented pectoral fins. Until recently they were only known by one genus, *Sinolepis*, from the Late Devonian of China. Several other sinolepidoids have since been recognised from numerous sites of Early–Late Devonian age in China and northern Vietnam. Sinolepidoids resemble the more primitive yunnanolepidoids in their head structure but have reduced trunk shields and long, segmented pectoral appendages. Recently a sinolepid was discovered near Grenfell, New South Wales, Australia, and has been named *Grenfellaspis*. This provides strong evidence that the Chinese and Australian terranes came into close proximity, enabling migration of fish faunas, during the Late Devonian.

The most successful groups of placoderms were the bothrolepidoids and the asterolepidoids, groups that flourished in the Middle and Late Devonian around the world. Asterolepidoids had long trunk shields with small heads, and robust, short, segmented pectoral appendages. The best known examples are *Asterolepis* (meaning "star scale"), known from Europe, Greenland and North America, and *Remigolepis* (from the Greek meaning "oarsman"), known from Greenland, China and Australia. *Remigolepis* is particularly unusual in having a short, stout pectoral appendage that lacks a joint. Many other asterolepids are known from around the world, such as the Middle Devonian genera *Pterichthyodes* from the Old Red

John A. Long

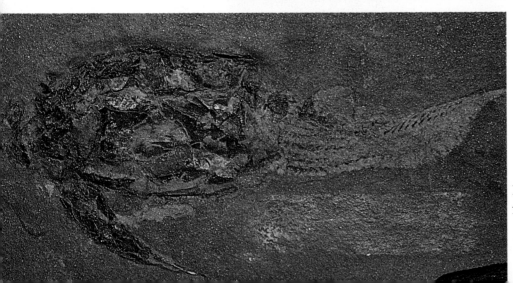

John A. Long

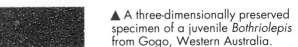

▲ A three-dimensionally preserved specimen of a juvenile *Bothriolepis* from Gogo, Western Australia.

►Reconstruction of one of the Antarctic species of *Bothriolepis* (*B. askinae*) from the Middle Devonian Aztec Siltstone of south Victoria Land.

◄ *Pterichthyodes*, one of the common Middle Devonian antiarchs found in the Old Red Sandstone sites of Scotland.

Evolutionary trends in antiarchs. The Early Devonian forms began with short, unsegmented pectoral fins, whilst the most advanced forms, the bothriolepids, had long, segmented pectoral fins, extending beyond the trunk shield.

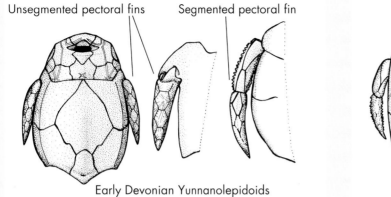

Unsegmented pectoral fins   Segmented pectoral fin

Early Devonian Yunnanolepidoids

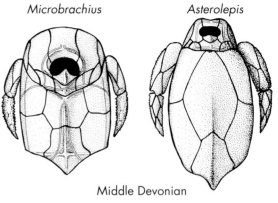

Microbrachius   *Asterolepis*

Middle Devonian

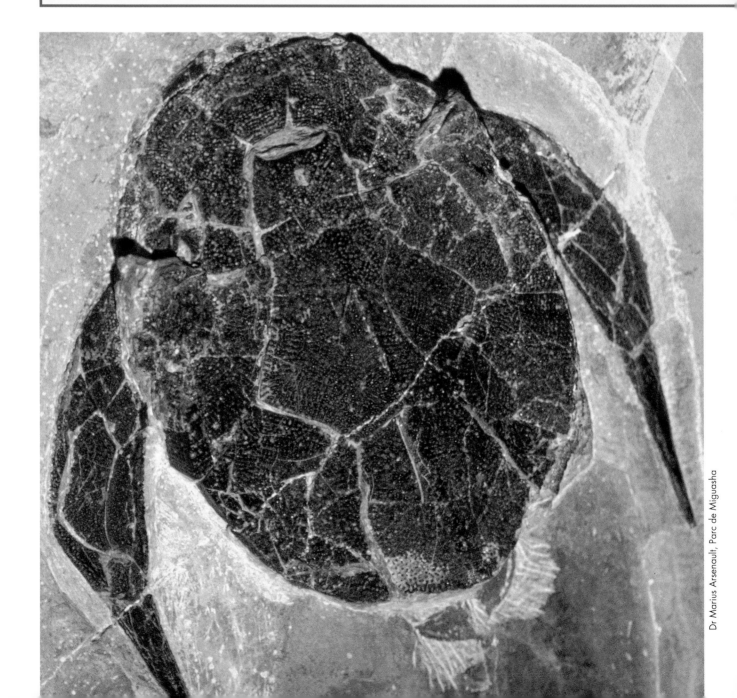

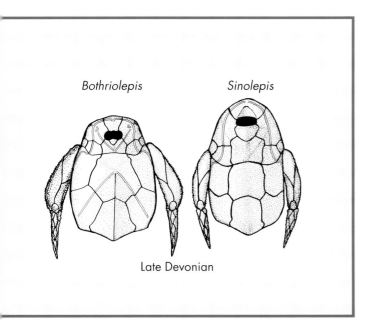

Bothriolepis

Sinolepis

Late Devonian

John A. Long

▲ Armour in side view of *Bothriolepis gippslandiensis*, from the Late Devonian of Victoria, Australia. This species has a low crest developed on its trunk shield. The photo is of a whitened latex peel.

◄ *Bothriolepis canadensis*, an advanced antiarch with long, segmented pectoral appendages, from the Late Devonian of Quebec.

Sandstone of Scotland, and *Sherbonaspis* from Australia. The asterolepids first appeared in marine environments and soon after invaded freshwater river and lake systems. By the end of the Devonian they were pushed out of the highly competitive marine realm and became freshwater river and lake dwellers.

The most successful placoderm of all time was undoubtedly *Bothriolepis* (meaning "pitted scale"), a little antiarch having long, segmented arms, and known from more than 100 species found in Middle and Late Devonian rocks of every continent, including Antarctica. *Bothriolepis* had a special feature of its skull that may have been the key to its success: a separate partition of bone below the opening for the eyes and nostrils, enclosing the nasal capsules, called a preorbital recess. Serially sectioned specimens of *Bothriolepis* show that inside the armour the fish had paired lung-like organs and a spiral intestine, preserved full of organic sediment, differing from the sediment type surrounding the fossil. *Bothriolepis* was probably a mud-grubber that ingested organic-rich mud for its food. Its long pectoral appendages could also have been used to push itself deeper into the mud for feeding. Another suggestion is that the fish used its long arms to walk out of the water and its "lungs" to breathe as it crawled out of the water to invade new pools free from predators and full of rotting vegetation. *Bothriolepis* is known mostly from freshwater deposits, as well as rarer marine sites such as the Devonian reefs at Gogo, Western Australia. Its ability to disperse around the Devonian world was via shallow seaways, from which it could invade river systems. Perhaps, like many modern fishes, it spent a large part of its life in the sea, moving upstream to breed and die. The fossil record suggests that most *Bothriolepis* species, irrespective of where they lived, died in freshwater habitats.

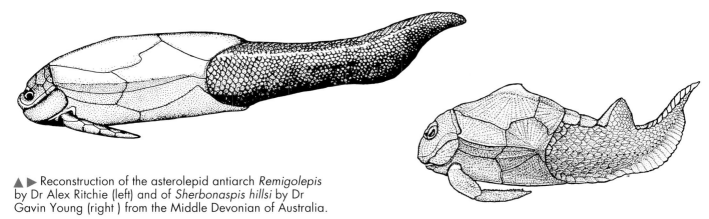

▲ ▶ Reconstruction of the asterolepid antiarch *Remigolepis* by Dr Alex Ritchie (left) and of *Sherbonaspis hillsi* by Dr Gavin Young (right) from the Middle Devonian of Australia.

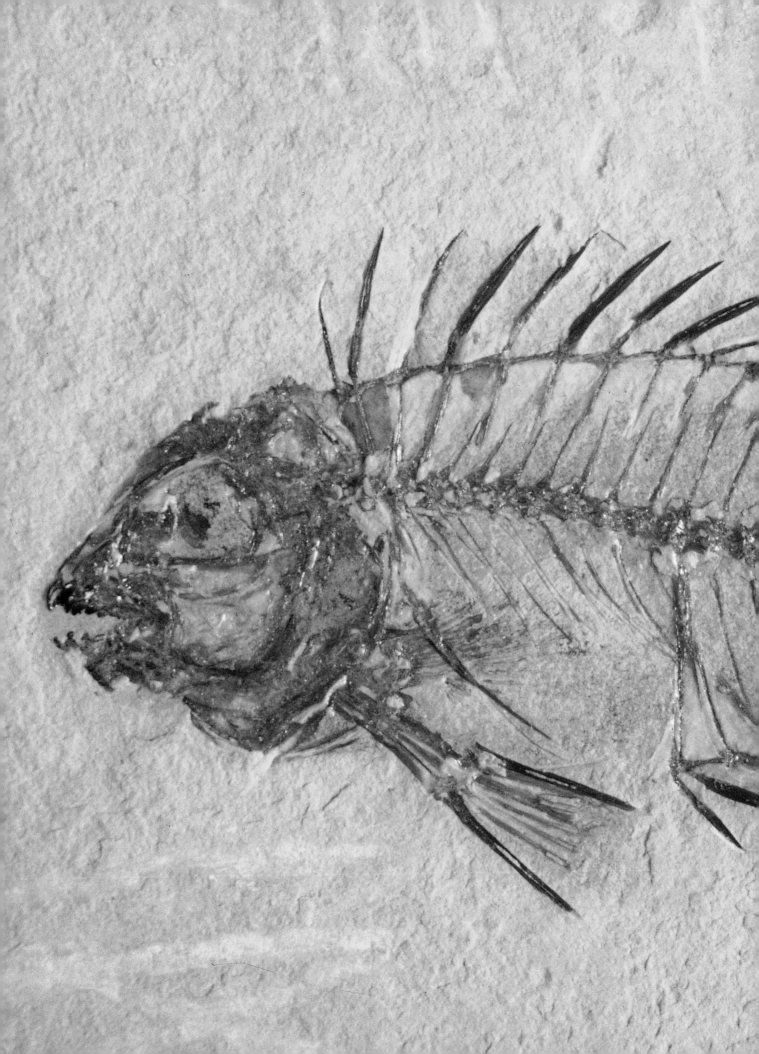

Chapter 6

TRUE
BONY
FISHES AND
THEIR
ORIGINS

CLASS OSTEICHTHYES

The osteichthyans (meaning "bony fishes") today form the largest and most diverse group of vertebrates and are represented chiefly by the ray-finned fishes (Subclass Actinopterygii). However, their origins date back to the Late Silurian, some 410 million years ago, when they were only a minor component of the ancient fish faunas. By the start of the Devonian Period all major groups of bony fishes had appeared, and before the end of the Devonian the first land animals had evolved from within the lobe-finned bony fishes. Buried within the early evolution of osteichthyan fishes lies the key to the most complex evolutionary transition in vertebrate history: how a water-breathing fish became a land-living amphibian. One of the secrets of the success of the "bony fishes" lies in their swim-bladder, an internal organ of buoyancy, which was to become modified into the lungs of land animals. The three major groups of bony fishes are the ray-fins (Actinopterygii), the lungfishes (Dipnoi) and the predatory lobe-fins (Crossopterygii). Aside from their differing physical features, especially their unique patterns of bones forming the skull and shoulder girdles, each group evolved unique tooth tissue types for specific feeding adaptations.

Today when we think of fishes in general terms, we think of the osteichthyans, the bony fishes. They are the major group containing the vast majority of the 23,000 or more living species of fish. But in contrast to their widespread success in today's seas, rivers and lakes, the earliest osteichthyans had a real battle to survive among the shoals of primitive Devonian predatory fishes. The first osteichthyans are very poorly known from fossils, represented by a few scales and mere fragments of broken bone. The oldest articulated remains, showing what their bodies and heads were like, are about 400 million years old. At this time the placoderms and acanthodians were reaching a peak of diversity, and sharks and osteichthyans swam quietly in the background.

As the Devonian progressed, the osteichthyans diversified into many different groups, mostly of lungfishes and lobe-finned predators, and these flourished right to the end of that period, with a few groups surviving a bit longer in geological time. Three genera of lungfishes still survive on three different continents (see Chapter 8), but only a single species of lobe-fin, the coelacanth, *Latimeria chalumnae* (see Chapter 9).

In this chapter we shall look at the basic architecture of an osteichthyan fish, together with the enigmatic remains of the most primitive fossil members of the group. The evolutionary success of the bony fishes lies not only in their flexible skeletal patterns, which allowed for a range of different feeding styles to evolve, but also in the nature of the varied tissue types developed within their skeletons. The meaning of the words "bone" and "teeth" will never be the same after one sees the range of tissue architecture that the various osteichthyans achieved, giving more strength to crushing plates, lightness and strength to principal skeletal components, and reinforcement to large stabbing fangs for prey capture. Furthermore, unlike the placoderms, sharks

Previous page: A well ossified internal skeleton, like that of this Eocene ray-finned fish from Monte Bolca, Italy, was one of the key features that made the osteichthyans a great success. This was largely due to the improvement in swimming efficiency resulting from having a strong vertebral column and ribs for tail muscle attachment. *L. Sorbini, Verona*

Inset: Teeth covered with layers of true enamel are a unique feature of osteichthyans and their derivatives, the higher vertebrates. *John A. Long*

BASIC STRUCTURE OF OSTEICHTHYANS

▲ A stained specimen of the sturgeon *Scaphirhynchus* showing internal cartilage and bones.

Osteichthyan fishes are characterised by having a well-ossified internal bony skeleton, although the earliest fossil forms show the least degree of ossification of the vertebrae and internal bones. Endochondral bone, formed around a cartilage precursor, is present, and marginal teeth are developed on the upper and lower jaw bones, specifically the maxilla, premaxilla and dentary. The premaxilla bears part of the infraorbital sensory-line canal. The gill arches are highly evolved relative to other jawed fishes, in that

pharyngobranchial elements are present with suprapharyngobranchials on the first two arches, and the hyoid arch has interhyal and hypohyal bones. The first two gill arches articulate to the same ventral median bone (basibranchial), which often bears small toothed bones to help crush up the food. Gular plates cover the underneath of the head, and a subopercular bone is present below the large opercular bone.

The pattern of bones in the shoulder girdle of osteichthyans is unique: a large cleithrum and clavicle make up most of the girdle, with smaller supracleithrum, post-temporal and postcleithrum or anocleithrum bones. This rigid girdle evolved in parallel to that of the placoderms to give support for the pectoral fin musculature.

The scales of osteichthyan fishes articulate and overlap with one another, more so than for any other fish; and in advanced osteichthyan groups, such as the higher groups of sarcopterygians and later actinopterygians, each rounded scale may have more than 75 per cent of its surface overlapped by neighbouring scales. This gives a high degree of rigidity to the trunk, which improves the efficiency of the swimming power.

The internal anatomy of an osteichthyan is unique among fishes in the presence of a swim-bladder (or lung) that enabled the fish to regulate its depth in the water column (and ultimately to breathe air). The musculature is also highly specialised in the presence of branchial levator muscles, interarcual muscles and transversi ventrali muscles.

Although these features are found in all members of the primitive osteichthyan groups, the great explosion of actinopterygians has seen some of these features highly modified or secondarily lost in special cases. Furthermore, many of these characters apply also to primitive fossil amphibians, again emphasising that these are really just an advanced lineage of osteichthyans with digits instead of fins. This fascinating transition is explored in detail in the final chapter of the book.

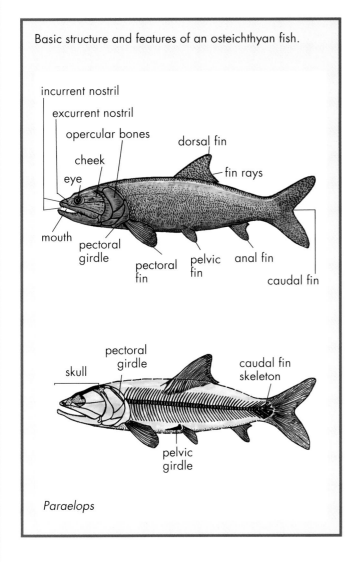

Basic structure and features of an osteichthyan fish.

incurrent nostril
excurrent nostril
opercular bones
cheek
eye
mouth
pectoral girdle
pectoral fin
dorsal fin
fin rays
pelvic fin
anal fin
caudal fin

skull
pectoral girdle
caudal fin skeleton
pelvic girdle

Paraelops

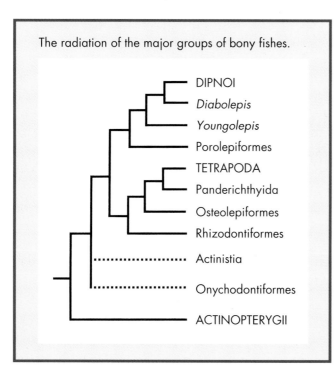

▲ Gurnard Perch, *Neosebastes pandus*, shows the typical features of the head of an osteichthyan. The gill arches are covered by a single large bone, the operculum, which opens and closes rhythmically as the fish respires. The mouth of this perch is actually a highly complex protrusable mechanism, a feature of more advanced ray-finned fishes.

The radiation of the major groups of bony fishes.

- DIPNOI
- *Diabolepis*
- *Youngolepis*
- Porolepiformes
- TETRAPODA
- Panderichthyida
- Osteolepiformes
- Rhizodontiformes
- Actinistia
- Onychodontiformes
- ACTINOPTERYGII

and acanthodians, the osteichthyans achieved one great evolutionary novelty that was to be the harbinger of a later successful invasion of land: an internal swim-bladder that could easily become modified as a lung.

The advantage of the swim-bladder is that it functions as a hydrostatic organ enabling the fish to rise or fall within the water column by regulating gases inside the organ. This means that energy is saved by not having to push forwards to develop lift with large wing-like pectoral fins (as in sharks and placoderms); thus the fins became freed to adapt for improved manoeuvrability within the water. With this came the great diversity of body shapes achieved by later lineages of ray-finned fishes, such as the deep-bodied palaeoniscoids (for example *Ebenaqua* and *Cleithrolepis*, mentioned in Chapter 7). In many living species of ray-finned fishes the swim-bladder can function as a temporary 'lung', enabling some species to exist outside the water for brief times. Mudskippers, small fishes that live among the mangroves and are able to climb trees, are great examples of this ability. The evolutionary transition from a swim-bladder, which is in effect a gaseous exchange organ, to a lung, also a gaseous exchange organ, was therefore not such a

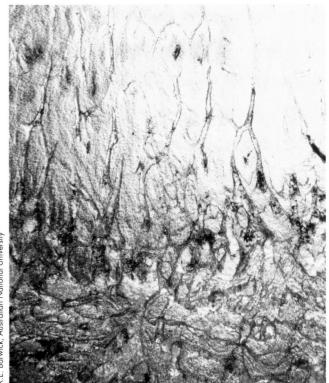

Cosmine, a primitive vertebrate tissue found only in sarcopterygian fishes. The function of cosmine may have been to support a complex surface sensory system.

outer surface skin (epithelium)

pore

flask

dentine

dentine canal

sensory cells

osteocyte space

bone

transverse canals

Clay Bryce, Western Australian Museum

R.E. Barwick, Australian National University

▲ Thin-section showing the dense mineralised dentine tissues making up the *Chirodipterus* tooth-plates.

◀ A common sea-dragon, *Phyllopteryx taeniolatus*, one of the many extraordinary examples of how the bony fishes, or osteichthyans, radiated into many distinctive forms and occupied a great variety of niches. The bony fishes today constitute the largest and most successful group of fishes, although the vast majority belong to one major group, the ray-finned fishes. The Devonian Period saw the emergence of the bony fishes with many, varied groups evolving and enjoying a short period of radiation prior to sudden extinction.

Scales of the earliest ancestors of the bony fishes belong to lophosteiforms. They have the basic shape and tissue stucture of the primitive osteichthyan scale but lack the typical peg and socket type articulation. Length under 2 mm.

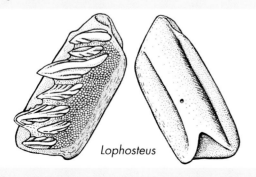

Lophosteus

▲ *Undina pencillata*, a fossil coelacanth from the Jurassic of Malm, Germany, shows the characteristic lobed pectoral and pelvic fins, spiny dorsal and anal fins, and a tufted tail fin.

▶ Palate of the Devonian lungfish *Chirodipterus*, from Gogo, Western Australia, showing the tooth-plates used to crush hard-shelled food. Many unique tissue types evolved within the teeth of osteichthyans in order to cope with a great variety of ways of capturing prey and reducing the food once in the mouth.

Dr K. Frickhinger, Munich

Enigmatic primitive sarcopterygians, like *Diabolepis* from the Early Devonian of China, represent forms intermediate between the major groups of lobe-finned fishes. The skull of *Diabolepis* is shown below.

skull roof

- pores of supraorbital sensory canal
- elevation of skull roof above pineal body
- anterior pitline
- middle pitline
- posterior pitline

- premaxillary
- anterior external nasal opening
- posterior nasal opening
- vomerine area
- parasphenoid
- articular area for palatoquadrate
- buccohypophyseal opening

palate

▼ The coelacanth, *Latimeria*, is the only survivor of the once great radiation of predatory lobe-finned bony fishes (crossopterygians). This specimen, on display in the Australian Museum, shows the reflection of the photographer, a human. The crossopterygians are the group which gave rise to land-living amphibians, and consequently, to mammals like us. A photo of a living coelacanth in its natural habitat can be seen in Chapter 9.

John A. Long

complex step but one that involves expansion of the surface areas within the organ to enable enough oxygen to be extracted from the air to support the organism.

The skeletons of osteichthyan fishes follow a conservative plan at their earliest origins, and most groups retained a similar pattern throughout their evolution. The notable exception is the ray-finned fishes, which underwent a great evolutionary radiation after the Devonian and kept on diversifying until the present day. The freeing up of their cheek and jaw bones enabled them to adapt to a great variety of specialised feeding mechanisms—from those of simple predators, to algal grazers, to tube-mouthed seahorses that suck plankton from the seawater ... (the variety is almost endless).

The structure of the most primitive osteichthyan fishes is not well known, unless we use one of the already-established groups such as the lungfishes or the lobe-finned crossopterygians. The basic features of these groups are shown in detail at the start of the chapters dealing with these groups. Here let us look at the basic osteichthyan features and tissues, pertaining largely to the stem-group osteichthyans or enigmatic fossil forms, such as the lophosteiforms, believed to be ancestral to all later bony fishes.

LOPHOSTEIFORMS:
ENIGMATIC ANCESTORS OF BONY FISHES

There are only two known genera of lophosteiforms, *Andreolepis* and *Lophosteus*, both of
which have been found in the Late Silurian deposits of
Oesel, an island in the Baltic Sea, off the coast of
Sweden. The lophosteiforms were first described by
the famous German palaeontologist Walter Gross in
1969 and are an enigmatic group known only from
isolated scales, teeth and some rare fragmentary
bones. They appear to be similar in structure to
primitive ray-finned fishes, although the teeth lack a
dense hard tissue called acrodin,
which mineralises on the tips of teeth
in nearly all ray-finned fishes.

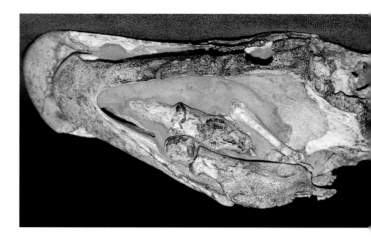

The tooth shape is conical with a
large central pulp cavity, and the teeth
were probably fused into the jaw,
lacking a root system. The scales have
a similar shape to the rectangular or
rhombic scales of early ray-finned
fishes but lack the layers of mineralised
shiny ganoine that make up the surface
ornament in the ray-finned fishes. The
shoulder girdle, known from a partial
cleithrum, is also of similar form to
that in the early ray-fin *Cheirolepis*. These tantalising
pieces give us a sketchy idea of the lophosteiforms as
looking a bit like the earliest ray-finned fishes, but
lacking the important tissue types that were to
characterise the group and be a foundation for later
major evolutionary changes in their skeletons.

John A. Long

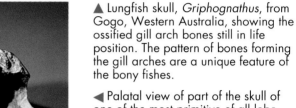

▲ Lungfish skull, *Griphognathus*, from
Gogo, Western Australia, showing the
ossified gill arch bones still in life
position. The pattern of bones forming
the gill arches are a unique feature of
the bony fishes.

◄ Palatal view of part of the skull of
one of the most primitive of all lobe-
finned fishes, *Youngolepis*, from the
Early Devonian of China.

▼ Western Blue devil, *Paraplesiops
meleagris*, shows the rich colour
patterns that some of the ray-finned
fishes evolved, and not seen to such
an extent in any of the other fishes
such as the sharks or jawless fishes.

THE OLDEST TRUE OSTEICHTHYANS

Scales belonging to primitive ray-finned fishes in Late
Silurian deposits of China and Russia have been
described quite recently. These scales have ganoine
layers and a well-developed peg for interlocking the
next scale in the row, and thus are accepted as
representing true actinopterygian scales (see Chapter
7). However, the first complete remains of ray-finned
fishes are not known until the Middle Devonian, some
35 million years later.

The oldest remains of lobe-finned fishes belong to
Early Devonian species from Arctic Canada,
Spitzbergen and China. These fossils consist of skulls
belonging to a group that had already achieved its basic
body plan and skull pattern—the porolepiforms (see
Chapter 9 for details), as well as some peculiar forms
that appear to be intermediate in structure between the

lungfish and other crossopterygian groups. Three
enigmatic genera are included in this lot: the possible
ancestors of the porolepiforms *Youngolepis* and
Powichthys, and the "proto-lungfish" *Diabolepis* (meaning
"devil scale"). These fishes are still poorly known and
their relationships within the osteichthyans has been the
subject of debate. Some workers argue that they are
closely allied to the porolepiforms and that lungfish and
porolepiforms are closely related sister-groups. Others
believe that the lungfishes are distinct from all other

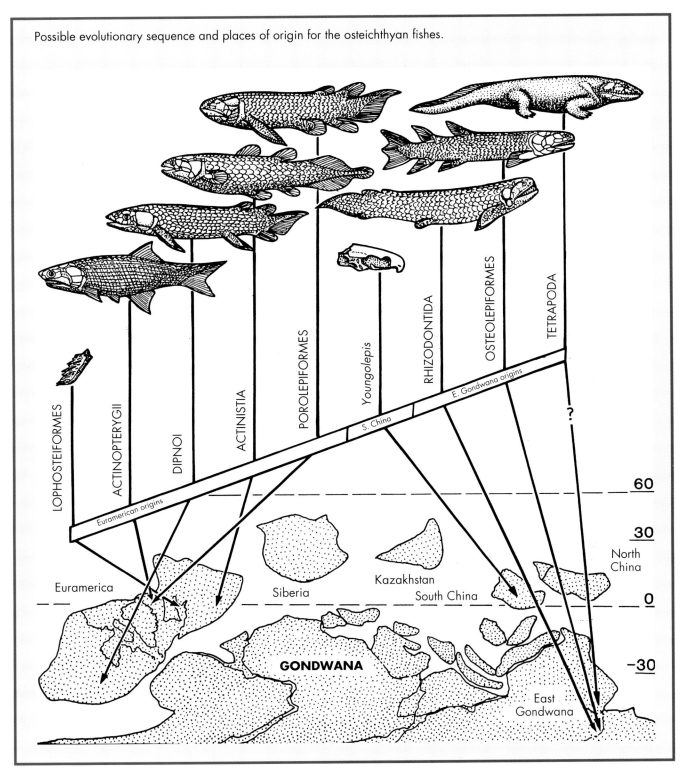

Possible evolutionary sequence and places of origin for the osteichthyan fishes.

osteichthyans and that the crossopterygians form a natural monophyletic group—in other words, that they share unique features that exclude lungfishes, such as the presence of many submandibular bones, or plicidentine tissue in the large teeth. At this stage we require more complete fossil material to be sure of the placement of these primitive forms and can at best place them as "stem-group" or basal crossopterygians.

The current consensus of opinion among fossil fish workers is that because *Youngolepis* and *Powichthys* possess an enameloid layer that dips into the pores of the cosmine, they are considered to be allied to the porolepiform crossopterygians. *Diabolepis* is generally regarded as more closely related to the lungfish than to any other group. However, whether lungfish and porolepiforms are closely related is still a controversial issue, and one that will be discussed in more detail in Chapter 9.

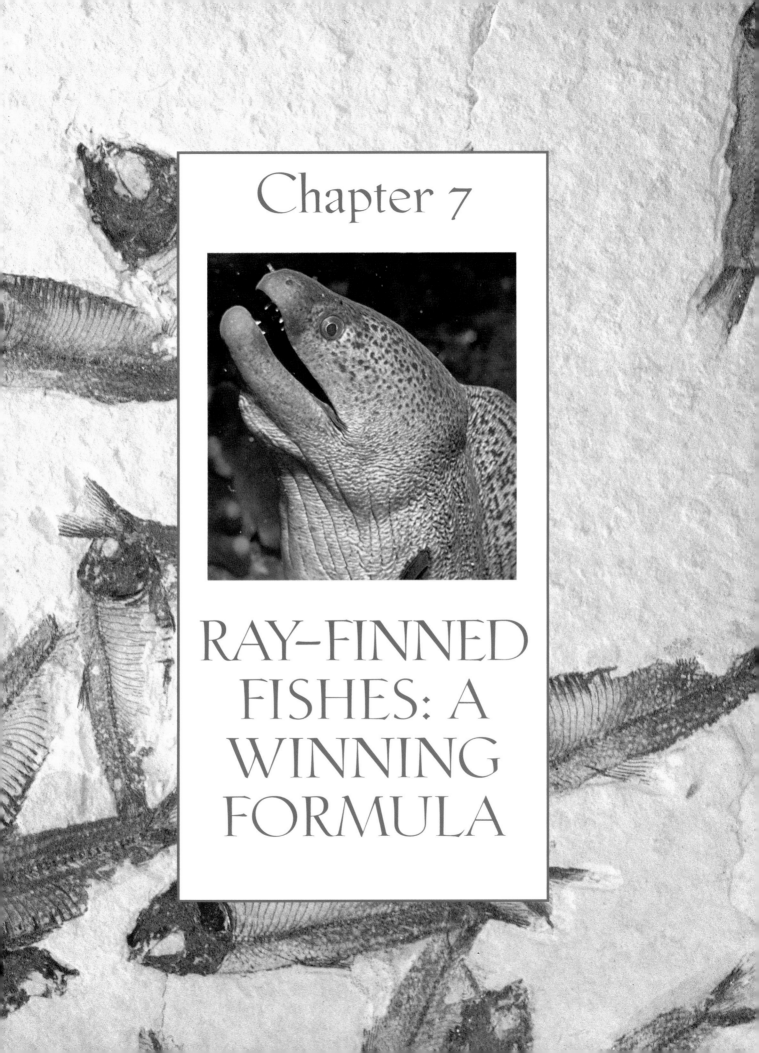

Chapter 7

RAY-FINNED FISHES: A WINNING FORMULA

SUBCLASS ACTINOPTERYGII

The ray-finned fishes (Actinopterygii) are the largest group of living fishes, comprising at least 23,000 species. The oldest ray-fins appeared in the Late Silurian, some 410 million years ago, known only from scales and fragmentary bones and teeth. The first complete fossils of the group belong to *Cheirolepis,* a peculiar form with tiny scales and fleshy pectoral fin lobes from the Middle Devonian of Scotland. In the Late Devonian about six other ray-fins appeared, including species well preserved in the Gogo Formation of Western Australia, which reveal much information about the anatomy and structure of the first ray-finned fishes, a group generally referred to as "palaeoniscoids". Living representatives of this primitive radiation include the reedfish, *Polypterus,* of Africa's rivers and lakes. In the Late Palaeozoic the group underwent a great burst of diversification, with some 40 or more families appearing in the Carboniferous and Permian Periods. During the Mesozoic the ray-fins evolved into more efficient swimmers and feeders, having cheek bones decoupled from the toothed jaw bones and greatly improved tail skeletons—the latter feature characterising the teleosteans, fishes that became the most successful group of vertebrates to modern times. The earliest teleosts appeared in the Middle Triassic some 220 million years ago, and by the Cretaceous they were well established in both marine and freshwater environments. The largest of all fishes may have been a ray-finned fish called *Leedsichthys,* fossilised in the London Clay of Britain. This poorly known giant may have reached sizes in excess of 12 m long. The percomorphs, the largest of the living groups of teleosts, evolved by the Cretaceous Period. By the dawn of the Cenozoic Era many of the living families of fishes had appeared, and the hectic roller-coaster ride of actinopterygian evolution had finally begun slowing down.

Perhaps one of the earliest descriptions of ray-finned fishes was by the famous Greek philosopher Aristotle (384–322 BC), who wrote in his *Historia Animalium* that "the special characteristics of the true fishes consist in the branchiae [gills] and the fins, the majority having four fins, but the elongated ones, such as the eels, having only two. Some such as the *Muraena* [moray eel], lack fins altogether." Thus one of the oldest scientific names of an actinopterygian fish came into existence. Later in the Middle Ages of Europe, the Church sent scholars such as Frenchman Pierre Belon (1517–1564) to university, after which he published his tome *La nature et diversité des poissons* (The

Nature and Diversity of Fishes). At about the same time another Frenchman, Guillame Rondelet, became a professor at Montpellier University and a specialist in studying fish (taken to mean ray-finned fishes in

Previous page: A school of fossil ray-finned fishes, *Knightia,* from the Eocene Green River Shale, Wyoming, USA. The ray-fins, or actinopterygians, were the most successful of all fishes, diversifying rapidly after the decline of many groups at the end of the Devonian Period.
John A. Long

Inset: The moray eel *Muraena* was one of the first ray-finned fishes ever named and described, by Greek philosopher and naturalist Aristotle (384–322 BC).
Clay Bryce, Western Australian Museum

BASIC STRUCTURE OF PRIMITIVE RAY-FINNED FISHES

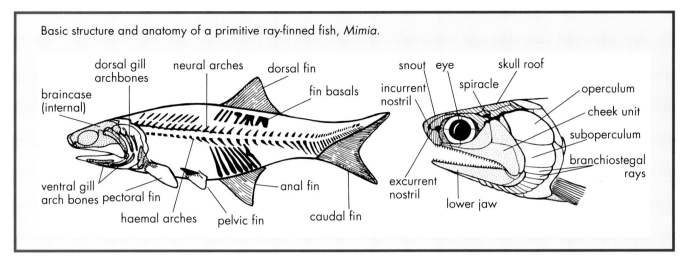

Basic structure and anatomy of a primitive ray-finned fish, *Mimia*.

The primitive Devonian actinopterygians are characterised by a long body with a single dorsal fin; all other early osteichthyans had two dorsal fins. Specialised features unique to actinopterygians are that the lower jaw bone has a large dentary bone (the one bearing teeth), which has a mandibular sensory-line canal enclosed within it; and that the jugal bone of the cheek has deep pit-line sensory organs.

The head of primitive ray-fins features a long gape with many sharp teeth set on the maxillary, premaxillary (upper jaws) and the dentary (lower jaw). The braincase is well ossified from several centres that are divided in maturity by ventral and occipital fissures. The midline of the palate has a long-toothed parasphenoid bone, and the inner surfaces of the jaws (both upper and lower) are covered by many smaller-toothed bones. The skull roof pattern has large frontal bones containing an open pineal foramen, paired parietals and nasals, a single large median rostral (or postrostral) bone, and the spiracular slit is still open in all Devonian and some subsequent species. The cheek is long, with a series of circumorbital bones carrying the infraorbital sensory-line canal, and there is a large preopercular bone above the maxillary. The operculogular series has a long opercular, large subopercular and a series of 12 or more branchiostegal rays. The shoulder girdle has a postcleithrum bone present, and the skeleton of the pectoral fin has a single row of internally ossified fin radials, as opposed to the longer arm-like skeletons in lobe-finned fishes. The fins are supported by many rows of segmented lepidotrichia and may have specialised "cut-water" scales on the fin's leading edges (fringing fulcra). The axial skeleton is poorly ossified in early ray-fins, consisting of perichondral shells of bone above the notochord, thus only neural and ventral arch elements are preserved with no tail skeleton ossifications. Later actinopterygians, like the successful teleosteans, evolved well-ossified bones in the tail.

The body scales had a well-developed peg-and-socket articulation to lock each scale into its rows, and had layers of a shiny surface tissue, ganoine, on dermal

▼ A reconstruction of *Cheirolepis*, the most primitive known ray-finned fish. Note the long body, small scales and long gape. The rigid cheek bones form an immovable plate over the jaw muscles, which in later forms became free, allowing a wide range of feeding mechanisms to evolve.

bones and scales. This type of scale is called the ganoid type and has layers of thin, laminar ganoine over a dentinous layer with vascular canals and a basal spongy bone layer. In later actinopterygians the ganoine is lost and the scales may take on rounded shapes. The teeth of all actinopterygians, apart from *Cheirolepis*, a very primitive form, have dense caps made of acrodin, a tissue formed of compact dentine. The external ornament of ray-finned fish bones is generally of fine parallel ridges and tubercles, each capped by shiny ganoine.

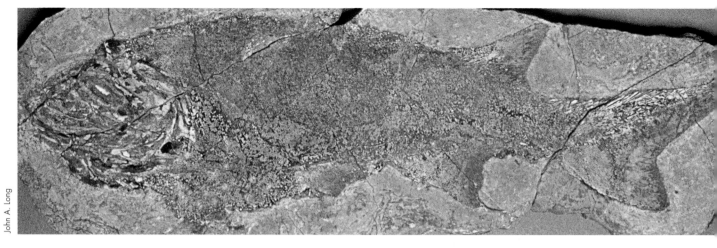

John A. Long

general). Rondelet's work *Universe Aquatalium Historiae* contained many fine descriptions of ray-finned fishes, but he also included anything else that lived in the sea including whales, crocodiles and shellfish.

The famous Swedish naturalist Carl von Linné (1707–1778), also known as Linnaeus, is considered the father of modern taxonomy. He was the first person to apply a binomial naming system to plants and animals. His classic work *Systema Naturae* published in 1758 was to set the stage for the great works describing and naming the diverse numbers of ray-finned fishes that were soon to be discovered. However, it was the famous Swiss scientist Louis Agassiz who, in his classic work of 1833–44 *Recherches sur les poissons fossiles*, first recognised, through comparison of fossil and living ray-finned fishes, the division of three main levels of organisation within the group—the Chondrostei, Holostei and Teleostei. Although Agassiz's classification of the Ganoidei gained much acceptance, his grouping contained many ray-fins as well as lungfishes and some crossopterygians and the odd acanthodian or two. The question of whether the ray-finned fishes formed a natural grouping was not really formalised until American palaeontologist Edward Drinker Cope erected his grouping in 1871, the Actinopteri, to distinguish the ray-fins from the lobe-finned fishes, and this was further set in concrete by Arthur Smith-Woodward, of the British Museum of Natural History, who coined the term Actinopterygii in 1891(being the equivalent of Cope's term). British biologist Goodrich followed Woodward's usage of Actinopterygii and it became adopted as the common name for this diverse group of living and fossil fishes.

▲ *Cheirolepis,* The earliest actinopterygian known from relatively complete remains. This specimen comes from the Middle Old Red Sandstone of Scotland (Middle Devonian age).

▼.Reconstruction of *Mimia,* a little fish less than 20 cm long that lived in the warm, tropical waters around the Late Devonian reefs of Gogo.

▼ Skipjack trevally, *Pseudocorax dentax,* an example of living ray-finned fishes. Today more than 23,000 species of ray-finned fishes have been described from the seas, rivers and lakes of the world, from hot volcanic springs to freezing Antarctic waters.

Clay Bryce, Western Australian Museum

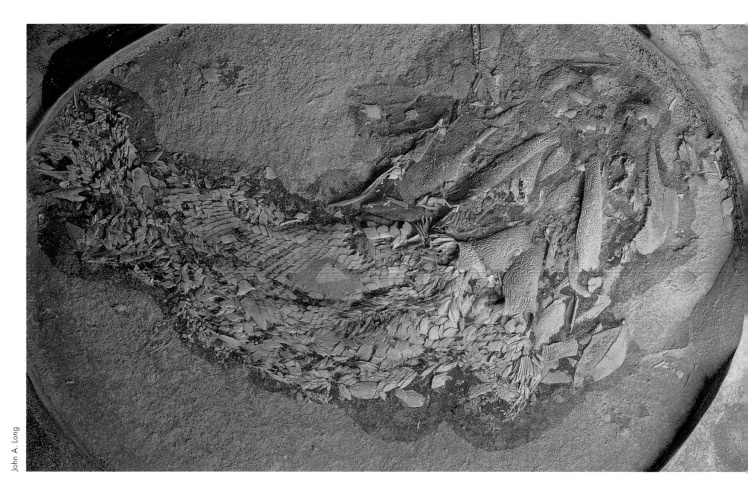

John A. Long

▲ *Mimia toombsi*, a Late Devonian actinopterygian superbly preserved from Gogo, Western Australia.

Today we know of at least 23,000 species of living actinopterygians, and new species are being discovered every year as nets bring up yet more unrecognisable fishes. The ray-fins technically belong in the Subclass Actinopterygii (literally meaning "ray wing"), as the fins of these fishes are supported by stiff bony spines or fin rays. The actinopterygians not only are the most diverse of all living vertebrates but also have penetrated the most hostile of all environments on Earth, from the deepest ocean depths ($-11,000$ m) to high mountain streams ($+4,500$ m), steaming hot volcanic springs ($43°C$) to freezing Antarctic waters ($-1.8°C$). They contain the smallest of all adult vertebrates, with one species reaching a mature size of only 7.5 mm, whereas the extinct ray-fin *Leedsichthys* may have grown to sizes in excess of 12 m.

The fossil record of actinopterygians is spectacular—from the oldest remains, scales of fish alive 410 million years ago, to complete three-dimensionally preserved Devonian species from the Gogo Formation of Western Australia and, generally speaking, an impressive variety of fossil species known throughout the rest of geological time right to the present day. As there are so many different families and groups within the Actinopterygii, this chapter will

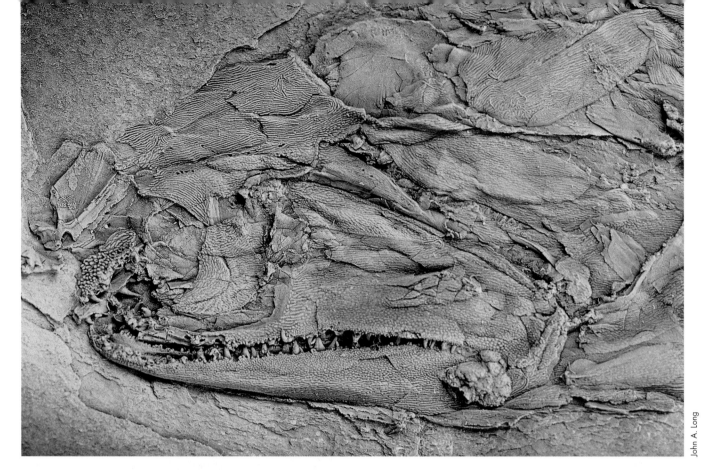

John A. Long

John A. Long

▲ Detail of the head of a primitive Devonian ray-finned fish *Howqualepis rostridens* from the Late Devonian Mount Howitt site, Victoria, Australia.

◄ *Polypterus*, the reedfish, lives in lakes and rivers of central Africa, and is considered to be a living representative of the most primitive known group of ray-finned fishes, the palaeoniscoids.

▼ A complete specimen of *Howqualepis*, from Mount Howitt, Victoria, showing the outline of the body and fins. These were freshwater fishes similar to modern day trout that probably hunted invertebrates and smaller fishes.

focus largely on the early origins and radiations of the group, with reference to major advances in their evolution that begat their most successful lineages. Many primitive forms of ray-finned fishes live today, as survivors from different stages in the progressive story of the group's evolution. It is largely from the anatomical studies of these primitive forms that the story of the actinopterygian evolution can be accurately reconstructed.

DEVONIAN RAY-FINNED FISHES

Scales of the ray-fin *Ligulalepis yunnanensis* date back to the Late Silurian of southern China, and similar scales have been found in Early Devonian limestones in Australia. The oldest skull roof of a ray-fin is of similar age from Arctic Russia, belonging to *Orvikuina*. This skull roof does not differ much from that of other

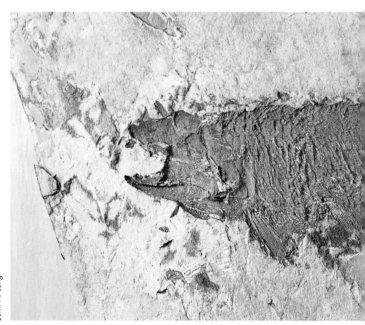

John A. Long

Dr Lance Grande, Field Museum Chicago

Devonian ray-fins, such as the Middle Devonian *Cheirolepis* and the Late Devonian *Mimia* and *Moythomasia* from Gogo in Western Australia. One primitive form from Australia, *Howqualepis*, was unusual in that it had teeth set into the median snout bone, the postrostral, separating the front of the upper jaw bones. In general the Devonian ray-fins are quite a homogeneous group, with little variation apart from those forms having a micromeric scale cover in which there are several scale rows to each body segment. Although most of the Devonian actinopterygians were fishes smaller than 15 cm or so, the largest, *Tegeolepis* from the Cleveland Shale of North America, reached nearly 1 m in length.

The main evolutionary trends seen in the Devonian ray-fins is seen in the shortening of the gape

▲ A fossil paddlefish *Crossopholis magnicaudatus* from the Eocene Green River Shale of Wyoming, North America. The paddlefishes are chondrosteans, like the sturgeons, and are primitive filter-feeders that use their gill arches to sift food.

▼ Reconstruction of *Mansfieldiscus*, a primitive river-dwelling Carboniferous palaeoniscoid from Mansfield, Victoria, Australia.

and enlargement of the opercular bones, the stabilisation of the skull roof as many small bones near the snout are replaced by a set pattern of large elements, and the change from the micromeric squamation to one of larger, rhombic scales. The braincase was apparently not well ossified in the most primitive forms such as *Cheirolepis*, but it was perichondrally and endochondrally ossified in Late Devonian forms found at Gogo. By the Late Devonian some genera were becoming widespread, such as *Moythomasia*, known from Western Australia and Europe.

There are two living representatives of this early actinopterygian radiation, the freshwater African reedfishes, *Polypterus* and *Calamoichthys*. These are elongated fishes with numerous spines forming the dorsal fin. In their early growth stages the cheek bones around the eye fuse with the upper jaw. They are very primitive in that unlike all other living and fossil ray-fins they lack several anatomical features, such as a surangular bone in the lower jaw, a pectoral fin internal skeleton with the leading bone perforated, and a hemipoetic organ above the medulla oblongata in the brain.

THE RISE OF THE PRIMITIVE RAY-FINS

The Early Carboniferous fish faunas of the world are dominated by the abundance of actinopterygians and sharks, although in most cases the latter are represented by only teeth and scales. The ray-fins are well preserved as whole-body fossils in many deposits of the Carboniferous and Permian Periods around the world, most notably from the sites near Edinburgh (Scotland), in the Bear Gulch Limestone of Montana (USA) from sites in Germany and Russia, and in Victoria (Australia). More than 40 families of ray-fins, comprising hundreds of recorded species, are known from the Late Palaeozoic, all of which are generally lumped as "palaeoniscoids" or ray-fins of primitive organisation.

Some particularly interesting families with deep-bodied shapes and highly specialised feeding mechanisms began to emerge by the Late Palaeozoic,

▲ Part and counterpart of a *Cleithrolepis* specimen as found in the Triassic sandstones at Somersby, near Gosford, New South Wales, Australia.

◀ A closer view of *Polypterus* showing the muscular pectoral fin, a primitive characteristic of some ray-fins also seen in the Devonian *Cheirolepis*.

such as the platysomid group. One of these, *Ebenaqua ritchiei*, from the Permian of Queensland (Australia), is exquisitely preserved and known from hundreds of specimens. It had a very small mouth, large dorsal fin, and greatly reduced anal and pelvic fins. It most likely fed on algae or weeds in the coal swamps it inhabited.

These and many other palaeoniscoids were, until recently, grouped in the Chondrostei, along with the two families of living primitive ray-fins, the sturgeons (Acipenseridae) and paddlefishes (Polyodontidae). Today the term Chondrostei is used only to include

FISHES AND DINOSAURS: LIFE IN A MESOZOIC ECOSYSTEM

William Stout, Pasadena, USA

The great southern supercontinent of Gondwana began its breakup early in the Jurassic Period, and by the Early Cretaceous all the continents had rifted away, leaving Australia the last to separate from Antarctica. India remained as an island drifting north from Antarctica from the Jurassic until the late Cretaceous, when it rammed into Asia and began the Himalayan mountain-building. In the northern hemisphere there was extensive contact between Asia and North America throughout most of the Mesozoic, enabling faunas to freely migrate.

During the early Mesozoic the most common plants were the seed ferns (pteridosperms), regular ferns, cycads, ginkgos and podocarp conifers. The forests were largely green, as flowering plants did not appear until the late Early Cretaceous and were not well-established until the Late Cretaceous. There were no grasses yet, and the ground cover would have been largely small ferns, psilophytes, seed-ferns, mosses, lichens, worts and fungi, in different combinations.

By the Mesozoic the seas were full of invertebrates of modern aspect but representing groups that are now mainly extinct. There were many types of cnidarians (jellyfish, corals, probably anemones), sea-mosses (bryozoans), molluscs of many kinds (clams, snails), as well as worms and arthropods (shrimps, crabs, early lobsters, king-crabs). Groups that were prolific at this time but did not survive later extinctions included cephalopods like the coiled ammonites, some up to 2 m in diameter, the squid-like belemnites, and many familes of brachiopods (lamp-shells). On land there were insects of modern aspect, as well as spiders, mites, scorpions,

millipedes, centipedes and most things you would find crawling under rocks or soil today.

Well into the Mesozoic Era the dinosaurs became established as the dominant vertebrate life on land, both in terms of their size and their diversity. The largest land animal to have ever lived was probably a brachiosaur named *Ultrasaurus*, reaching to an estimated 30 m long and weighing up to 100 tonnes, although longer-necked sauropods such as *Seismosaurus* may have reached lengths of nearly 50 m. Large predators, such as *Tyrannosaurus*, grew to 16 m long, and hunted these gargantuan plant-eaters. The first mammals appeared at almost the same time as the first dinosaurs. These were small rat-sized creatures throughout most of the Mesozoic. Despite their uninteresting appearance, most of mammalian higher evolution took place in the Mesozoic, as seen by their teeth, resulting in the three main lines of mammals alive today: egg-laying monotremes, pouched marsupials and the eutherians, which give birth to well-developed young. The first bird, *Archaeopteryx*, had evolved by the Late Jurassic and more than 60 species of birds are known in the Cretaceous. In the seas the large reptilian predators were the ichthyosaurs, plesiosaurs and, late in the Mesozoic, the lizard-like mosasaurs.

Crocodiles, lizards and turtles were abundant, along with amphibians such as frogs and salamanders. The large labyrinthodont amphibians mostly died out by the Late Triassic, although some forms persisted as late as into the Early Cretaceous in Gondwana and parts of Asia. Fishes would have had many predators both in terrestrial and marine environments.

these two families, which both have good fossil records spanning back to the Late Cretaceous. These chondrosteans are characterised by having braincases that lack eye muscle insertion pits (myodomes); the upper jaw bones (premaxillae, maxillae and dermopalatines) are fused together; and the upper jaw cartilage meets in the midline at the front of the mouth. A paddlefish, *Crossopholis*, from the Eocene, indicates that the group has changed little in appearance over the past 40 million years.

There are many other groups of primitive ray-finned fishes that enjoyed a great radiation in the Late Palaeozoic and Early Mesozoic, such as the perleidiform, pholidopleuriform and redfieldiform fishes. These include a variety of fossil species that have been found in Permian, Triassic and Jurassic deposits around the world. Most of the well-preserved fishes from the Triassic of the Sydney Basin, Australia, fall into one of these groupings, such as *Dictopyge* (a redfieldiform), *Cleithrolepis* and *Manlietta* (perleidiforms).

NEOPTERYGII: THE DAWNING OF SUCCESS

By the Late Palaeozoic there were many successive higher levels of actinopterygian evolution. It was then that actinopterygians acquired a vertical suspensorium, where the lower jaws articulate with the upper jaws by a vertically oriented quadrate bone. Also the cheek bones that were firmly united with the

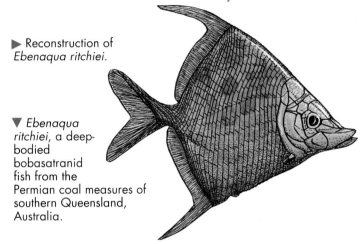

▶ Reconstruction of *Ebenaqua ritchiei*.

▼ *Ebenaqua ritchiei*, a deep-bodied bobasatranid fish from the Permian coal measures of southern Queensland, Australia.

John A. Long

▲ *Brookvalia gracilis*, a little ray-finned fish from the Triassic of the Sydney Basin, at Brookvale, New South Wales.

upper jaw in primitive forms became freed, allowing a great variety of feeding mechanisms to evolve. Many accessory bones evolved in the cheek region, giving endless patterns of bones in the heads of the different groups. However, the biggest innovations were in the fin and tail skeletons.

The advanced group of actinopterygians, called Neopterygii, have fin rays of the dorsal and anal fins equal in number to their support bones, and the upper jaw bones are fused in the midline. Inside the mouth they have well-developed pharyngeal tooth-plates that have been consolidated into stout bones for grinding up food as it passes into the back of the gullet. Within the Neopterygii are two main groups, the Ginglymodi, containing the garfishes, and the Halecostomi, containing bowfins and teleosteans.

TELEOSTEANS: THEIR EARLY BEGINNINGS

The great majority of living fishes are teleosteans, a group that has humble origins back at the dawn of the age of dinosaurs, in the Triassic Period. The key feature defining the teleosts is the presence of certain bones in the tail skeleton, called uroneurals, which function to stiffen the dorsal lobe of the tail and support a series of dorsal fin rays. This simple innovation gave the group greater swimming power and allowed a great variety of body shapes to evolve. But teleosts also have extra features that other fishes

John A. Long

John A. Long

▲ *Leptolepis talbragarensis*, from the Late Jurassic of New South Wales, one of the first of the advanced ray-fins (teleosteans).

▼ *Leptolepis koonwarriensis*, an Early Cretaceous species from Koonwarra, near Leongatha, Victoria.

lack. They underwent modifications to the jaws, making the front toothed bones of the upper jaw, the premaxillae, free to move independently of the main upper jaw bone, the maxilla. Their large ventral gill arch bones, the basibranchials, have unpaired tooth-plates on them, and they have many specialisations of the head musculature.

There are many large groupings of fishes within the Teleostei, but for simplicity's sake we shall here consider a few of the main fossil forms leading to the most advanced group, the percomorphs.

Many primitive fossil teleosts are known from the Mesozoic Era. *Leptolepis*, for example, a common genus known from many species around the world in Jurassic and Cretaceous freshwater deposits, has a trout-like appearance but when the head is examined in detail it shows many primitive features. The premaxilla is tiny and still tightly bound to the maxilla, and there are additional supramaxilla bones above the maxilla. Fishes like *Leptolepis* were highly successful, being able to migrate around the world and no doubt capable of living in both marine and freshwater environments, perhaps favouring the latter for breeding, like many living fishes such as the salmon.

Fish fossils showing the intermediate stages between the primitive ray-finned groups and the stages leading to early teleostean evolution are found from a famous locality in South America where limestone nodules wash out of eroding hills. The area is the

▲ *Wadeichthys oxyops*, another early teleost from the Koonwarra site, Victoria.

▶ Reconstruction of *Leptolepis koonwarriensis*, a primitive teleost from the Early Cretaceous lake deposits of Victoria, Australia.

Araripe Plateau near Jardim, northeastern Brazil, and like fossils from Gogo and central Queensland in Australia, the fish can be prepared by weak acetic acid to reveal extraordinary details of their skeletons. Many of the commonest fossils here, for example *Vinctifer* and *Rhacolepis,* are commonly sold by fossil dealers around the world.

The Ichthyodectiformes were another successful group of primitive teleosts, known largely from marine deposits of Cretaceous age. Some of these fishes reached large sizes, 5 m or more, and all were highly predatory fishes whose jaws bristled with sharp teeth. They were active, free-swimming hunters which may have preyed on other fishes as well as unwary smaller marine reptiles, including young ichthyosaurs and plesiosaurs. Large, well-preserved skulls of these have been found in central Queensland (for example *Cooyoo australis* and *Pachyrhizodus marathonensis*) and, when three-dimensionally prepared, they reveal magnificent details of the skulls

▲ The Koonwarra fish site in east Gippsland, Victoria, was reopened for further excavations in the early 1980s. The team excavating the site shown here include from the right, Tim Flannery, Alex Ritchie, Jim Warren (squatting), and Paula Kendall (left).

▶ Reconstruction of *Xiphactinus*, a large predatory teleost from the Cretaceous of North America.

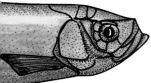

▼ Reconstruction of *Koonwarria manifrons*, a primitive teleost found at Koonwarra, Victoria.

fin and upper jaw musculature. These few innovations gave the group an ability to better manipulate and break down food inside the mouth by being able to retract and move forward the pharyngeal tooth-plates situated at the front of the gullet. Food was captured or bitten off by the mouth using the toothed bones of the upper and lower jaws, then crushed inside the gullet by the powerful grinding mills, the pharyngeal tooth-plates.

The most successful group of fishes within the Neoteleosts are the "spiny fins" or Acanthopterygii. These are characterised by one further refinement of the pharyngeal tooth-plates in having enlarged second and third epibranchial bones (in the gill arches) and insertion of the retractor dorsalis muscle onto just the third pharyngobranchial bone. In simple terms the

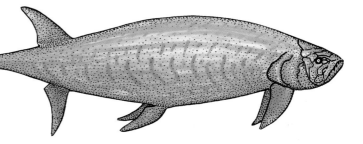

acanthopterygians fishes can protrude their jaws in a number of efficient ways, turning the mouth into a versatile prehensile device. There are 13 orders of fishes within this group, testifying to the success of their feeding mechanism. The most successful of the acanthopterygians are the Percomorpha, a poorly defined group including many of the living families of fishes. Within this are the perch-like fishes (Order Perciformes), comprising approximately 20 suborders, 150 families and nearly 7000 species of fish. (It is the largest order of vertebrates.) However, the Percomorpha group cannot be

▼ Reconstruction of *Uarbryichthys*, a Late Jurassic teleost from the Talbragar River fish beds of New South Wales.

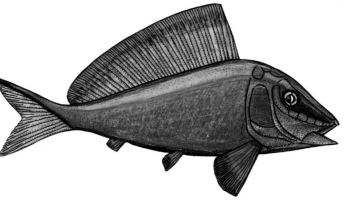

and braincases. One of the fishes from the same deposits is *Duggaldia*, representing a more advanced kind of teleost, possibly belonging to the Neoteleostei ("new teleosts").

Neoteleosts are characterised by features of the head musculature—having a retractor dorsalis muscle present which could operate the upper pharyngeal tooth-plates—and they all have a median rostral cartilage between the premaxillae and the braincase, as well as specialised features of their pectoral well defined and might represent a taxonomic "duffel bag" in which many similar-looking species are placed, without our really knowing if they are closely related. Nearly all of the major groups of living fishes have fossil representatives, either as well-preserved whole fishes or as just ear-stones (otoliths), which are readily preserved in marine sediments.

The famous Green River shales of Wyoming, North America, have yielded hundreds of different species of ray-finned fishes that lived in the Eocene Period about 40 million years ago, in a series of rivers and lakes. This window into the past shows us the whole evolutionary succession of ray-fins, from primitive chondrosteans and gars through to advanced neoteleosteans. Many of the fishes can be bought in mineral and fossil shops around the world, and nearly every museum in every country has fossil fishes from these deposits. Most commonly found are the little teleosts *Knightia* and *Diplomystus*.

Another well-known fossil site of Eocene age, but representing marine habitats, is located in the mountains of northern Italy near the village of Bolca. These superb fish fossils were collected as far back as the sixteenth century, as noted by Andrea Mattioli in 1555. A museum that included many fish fossils was set up in Verona by Francesco Calceolari in 1571. Today the Natural History Museum in Verona houses the finest collection of Eocene marine fish fossils ever

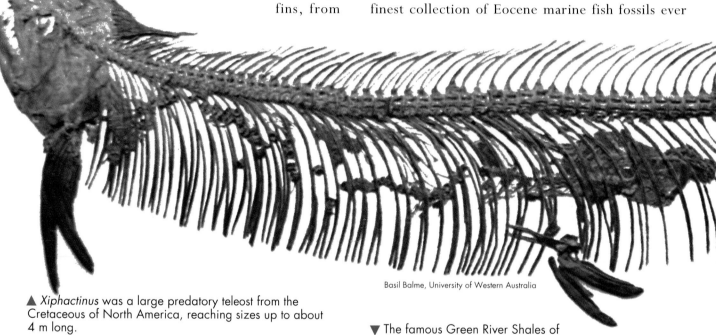

Basil Balme, University of Western Australia

▲ *Xiphactinus* was a large predatory teleost from the Cretaceous of North America, reaching sizes up to about 4 m long.

▼ The famous outcrops of Cretaceous marine Santana Formation, near Araripe, Brazil, have yielded an extraordinary fauna of fossil fishes, reptiles, insects and plants.

▼ The famous Green River Shales of Wyoming, USA, represent ancient lake deposits dating to about 40 million years old. Each year many thousands of fish fossils from these sites are mined and commercially sold around the world.

Dr John Maisey, American Museum of Natural History

Dr Lance Grande, Field Museum Chicago

assembled. Again, many different fossil species have been found, most representing families of living teleosts, as well as remarkably preserved fossil sharks and rays. So far more than 150 fish species have been recorded from the Monte Bolca sites, and much of the work describing these fishes has been done by Dr Lorenzo Sorbini, Curator at the Verona Natural History Museum. Examples of Monte Bolca fishes from extant families are the anglerfish *Lophius* and the flattened flounder *Eobothus*. Others like *Exellia* represent their own unique families.

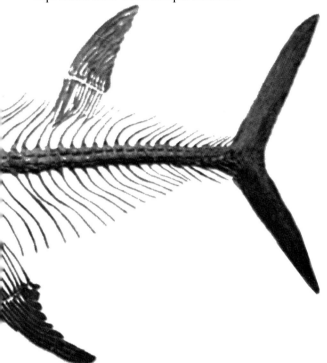

Dean Lee, Underwater World, Perth

▲ The Antarctic cod *Notothenia* lives in the seas around Antarctica, in temperatures close to −2°C. It has antifreeze enzymes in its blood to prevent it from freezing solid. Such fish demonstrate the extraordinary extent to which teleostean fishes have evolved to live in a great diversity of environments, from Antarctic seas to hot volcanic springs of 43°C.

▼ The head of *Tharrias*, an Early Cretaceous teleost from Brazil.

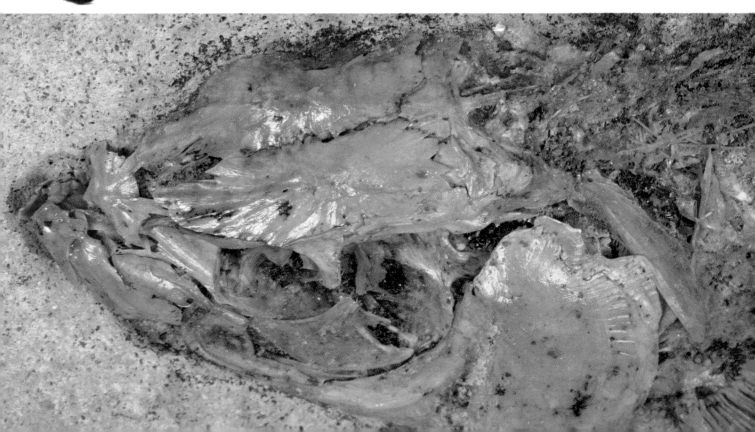

American Museum of Natural History

The tail skeleton of *Brannerion* shows the specialised uroneural bones that gave teleosts a powerful swimming ability.

epural 1

uroneurals 1–3

neural arch

hypural 6

preural neural arch

ural centrum 1

ural centrum

hypural 1

preural centrum

parhypural

▲ The tail skeleton of *Brannerion*, a Cretaceous teleost, shows the important specialisations that enabled the group to become so successful—enlarged tail bones called hypurals.

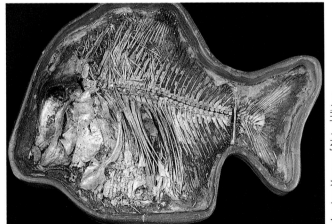

American Museum of Natural History

▲ *Neoproscinetes*, a deep-bodied teleost from the Santana Formation, Brazil. The specimen has been acid-prepared.

John A. Long

▲ *Diplomystus*, another common fish found in the Eocene Green River Shales of North America.

▼ The head of *Calamopleurus*, from the Early Cretaceous Santana Formation of Brazil. *Calamopleurus* was a fast-swimming predator.

▲ *Eobothus,* one of the soles, from the Eocene of Monte Bolca. Some fish families have changed very little in 40 million years.

▼ *Phaerodus testis*, an osteoglossid from the Eocene Green River Shales, Wyoming, USA.

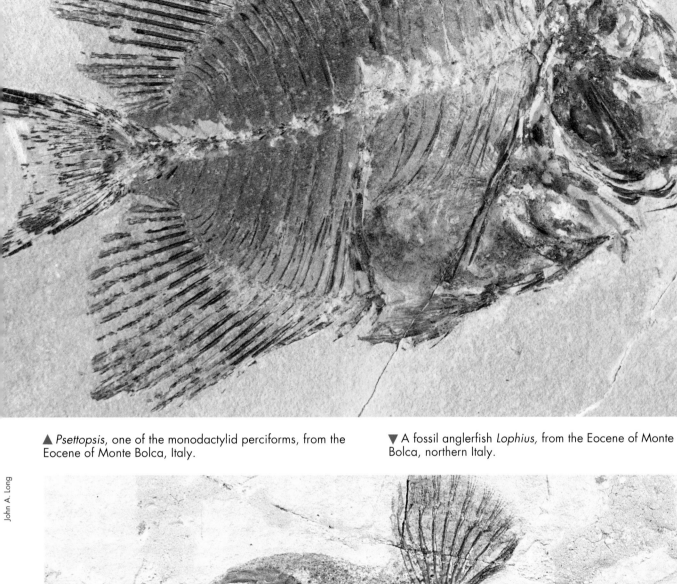

▲ *Psettopsis*, one of the monodactylid perciforms, from the Eocene of Monte Bolca, Italy.

▼ A fossil anglerfish *Lophius*, from the Eocene of Monte Bolca, northern Italy.

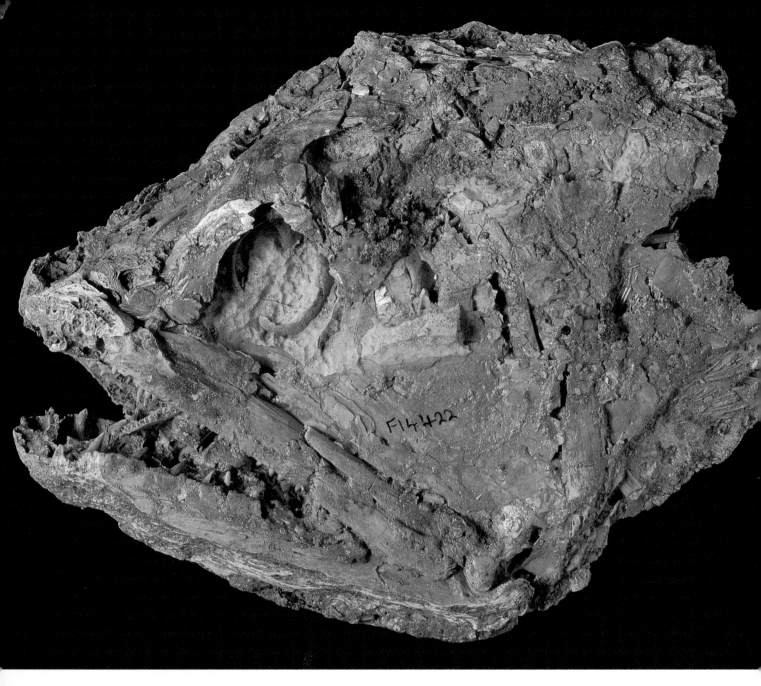

The head of *Cooyoo australis* from the Early Cretaceous marine deposits of Queensland.

◄ The famous Monte Bolca site, northern Italy, has yielded many species of Eocene marine fishes, both ray-fins and chondrichthyans. Here Dr Lorenzo Sorbini of the Natural History Museum in Verona is searching for fish fossils.

It is clear from the Eocene fossils found in Wyoming and Italy that many of the modern groups of fishes, represented at family level, had appeared by the early part of the Cenozoic Era. Most of the evidence for their early appearances come from the ear-stones (otoliths) found in marine sediments. Many hundreds of these can be extracted from a bulk sample of marine sediment using the correct sieving procedures, and they have been used around the world to identify the nature of Tertiary fish faunas from varying latitudes.

A selection of fossil teleost otoliths (ear stones) from the Miocene of Australia. All specimens are between 0.3 and 1 cm in length.

Platycephalus petilis, flathead

Gadus refertus

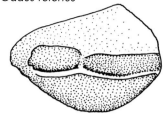

Coelorhynchus elevatus, whip-tail

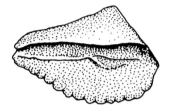

Sillago pliocaenica, whiting

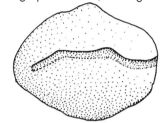

Heterenchelys regularis

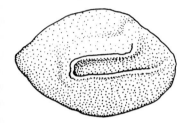

Pterothrissus pervetustus

Astroconger rostratus, eel

Lactarius tumulatus, cowfish

Monocentris sphaeroides

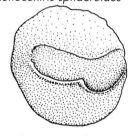

Merluccius fimbriatus

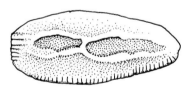

Trachichthyodes salebrosus

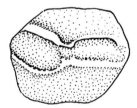

Ebastodes fossicostatus

Megalops lissa

Antigonia fornicata

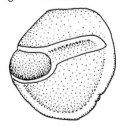

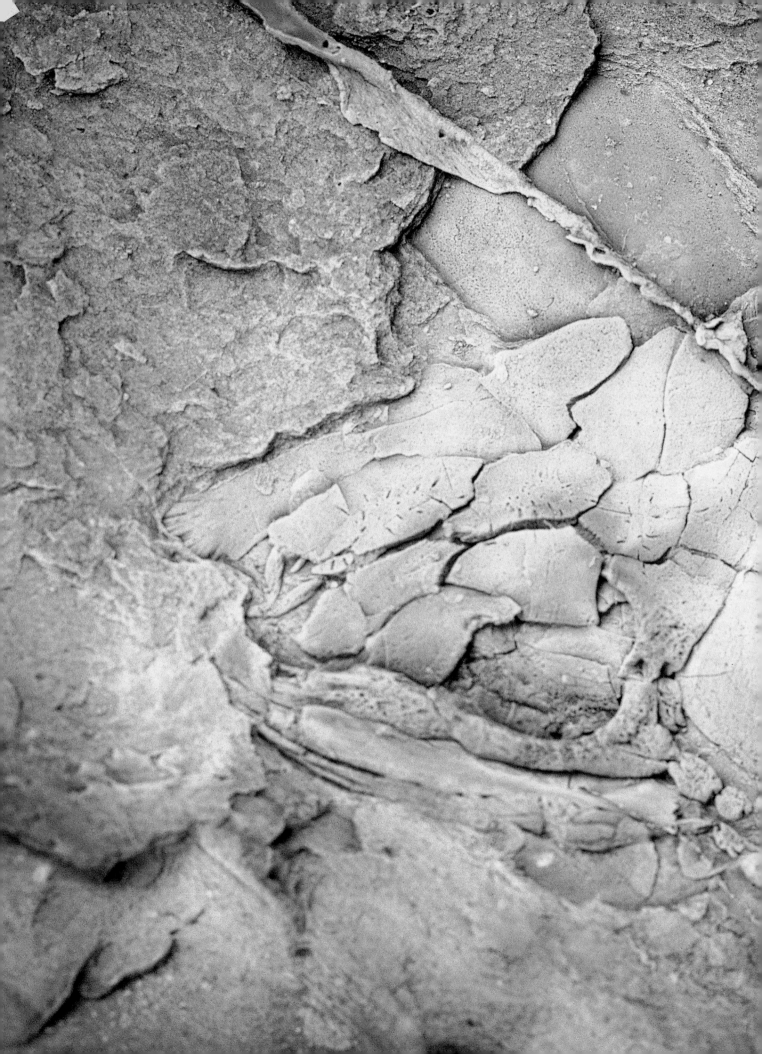

Chapter 8

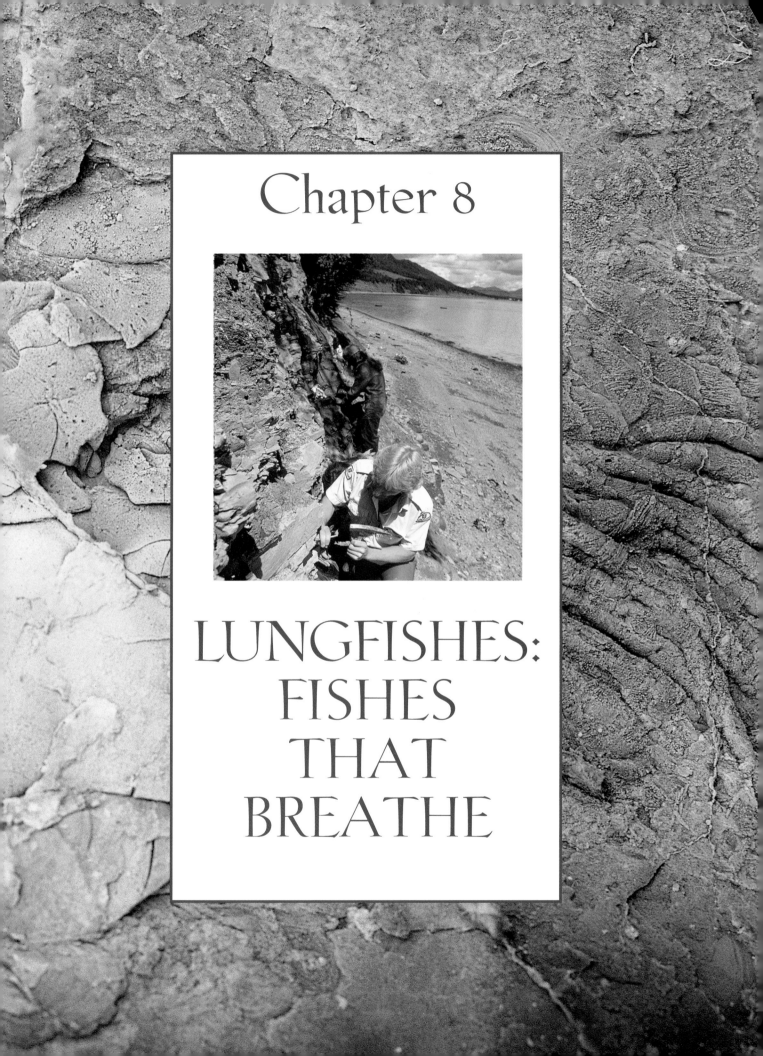

LUNGFISHES: FISHES THAT BREATHE

SUBCLASS DIPNOI

The lungfishes have an almost complete fossil record in Australia, including superb three-dimensional skulls of Devonian age from New South Wales and Western Australia. Although the first lungfishes, dating back to about 400 million years, lived in the sea, all lungfishes, since about 340 million years ago, inhabited freshwater environments. The fossil record tells us that lungfishes acquired the ability to breathe air independently of other vertebrates and are therefore not considered ancestral to the first four-legged land animals, the amphibians. Fossil burrows tell us that back in the Late Palaeozoic lungfishes could aestivate (lie dormant) enclosed in baked mud during the dry season, awaiting the next season's rains. The story of lungfish evolution is one of frantic and rapid change during the Devonian, the "dipnoan renaissance", with a much reduced rate of evolutionary change from the end of the Carboniferous Period to recent times. The three genera of living lungfishes occur in Africa, South America and Australia. The Queensland lungfish is the most primitive of these "living fossils" — fossils of the Australian lungfish indicate that this species has remained unchanged in Australia for at least 100 million years, making it the most enduring species of vertebrate known on Earth.

Dr Alex Ritchie, Australian Museum

▲ The Queensland lungfish *Neoceratodus forsteri*. When living lungfishes were first discovered, they were thought to be amphibians or reptiles, not fish.

Previous page The head of *Barwickia downunda*, shown from side view, with slight compaction of features.
Kristine Brimmell, Western Australian Museum

Inset The outcropping blue silty sandstones of the Escuminac Formation, near Miguasha, on the Gaspé Peninsula, Quebec, has yielded a superbly preserved fauna of Late Devonian fishes, including many fine specimens of lungfishes.
Dr Marius Arsenault, Parc de Miguashua

The dipnoan name for lungfishes comes from the ancient Greek meaning "two lungs" because of their ability to breathe twice, with gills in the water and with lungs to gulp air. Although fossil lungfishes have been known from the Old Red Sandstone rocks of Scotland for more than 200 years, when the three known living lungfishes were discovered some of the early zoologists could not believe that they were fishes.

The first discovery of a living lungfish was made in 1836 by Viennese naturalist Johann Natterer, who collected *Lepidosiren* specimens from the mouth of the Amazon River, Brazil. He sent his material to Leopold Fitzinger, Curator of Reptiles at the Imperial Museum in Vienna, who wrote to Count von Sternberg about the animal. The letter was later read before the Society of Natural History in Jena. Although Fitzinger's specimens were gutted, a remnant of the lung was found, and this, together with the unusual nostrils placed near the upper lip, led him to describe the creature as "undoubtedly a reptile". In the following year the great British anatomist Richard Owen was able to observe a specimen of an African lungfish collected from the Gambia River and presented to the Royal College of Surgeons. Owen, too, was puzzled by the unusual mosaic of fish and amphibian features seen in the specimen, but concluded that given the structure of the nasal sacs and nostrils it must be a fish. Although Owen had earlier proposed the genus name *Protopterus* for the specimen, he changed it back to *Lepidosiren* after receiving Natterer's paper of 1838. The first Australian lungfish *Neoceratodus* was found in Queensland in 1870, 32 years after the closely-related genus *Ceratodus* was described by Swiss palaeontologist Louis Agassiz from fossil tooth-plates. Gunther, of the British Museum of Natural History, gave the first detailed anatomical description of the Queensland lungfish in 1871 and recognised it as being a fish. Because *Neoceratodus* was immediately recognised as being the most primitive of the three genera of living lungfishes, most subsequent work focused on the anatomy and embryology of this genus.

Muller erected the name Dipnoi in 1844, but the confusion over how dipnoans, both living and fossil, should be classified resulted in a string of newly proposed names over the next two decades, none of which has stood the test of time: Order Ichthyosirenes (Castelnau 1855), Family Pneumoichthyes (Hyrtl 1845), Ichthyosirens (M'Donnel 1860), Order Protopteri (Owen 1853), Order Pseudoichthyes (Owen 1859). Hogg (1841) probably takes the cake for placing them in his Tribe Fimbribranchia and Family Amphibichthyidae! Lungfishes are now placed in the Subclass Dipnoi, and three living genera are placed in two families, the Lepidosirenidae (*Lepidosiren*, *Protopterus*) and the Ceratodontidae (*Neoceratodus*).

Today we have an excellent fossil record of lungfishes, especially so from Australia, where remains of these fishes come from almost every period of geological time since their first appearance at the beginning of the Devonian.

The remains of the first lungfishes come from limestones and shales containing abundant fossils of marine invertebrates, indicating that the lungfishes lived in shallow sea environments. Since the end of the Carboniferous Period dipnoans have inhabited only freshwater environments, and in the Permian Period some had acquired the ability to aestivate. The story of lungfish evolution can be summarised as one of frantic, marvellous change in the Devonian Period, "the dipnoan renaissance", followed by slow steady change during the later part of the Palaeozoic Era and into the Mesozoic Era. Since then they have remained almost

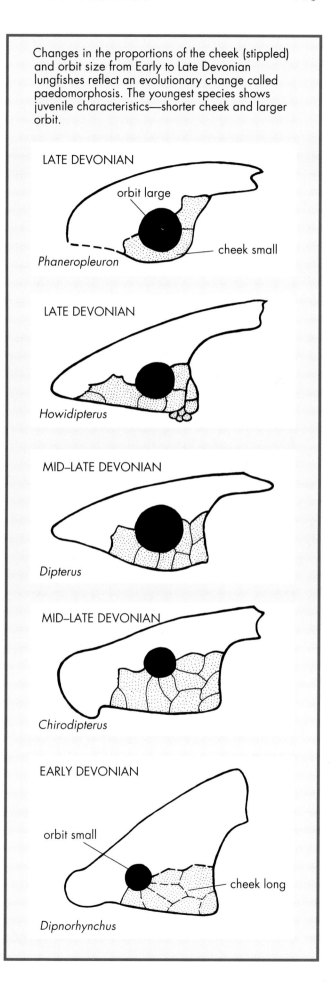

Changes in the proportions of the cheek (stippled) and orbit size from Early to Late Devonian lungfishes reflect an evolutionary change called paedomorphosis. The youngest species shows juvenile characteristics—shorter cheek and larger orbit.

LATE DEVONIAN

orbit large

cheek small

Phaneropleuron

LATE DEVONIAN

Howidipterus

MID–LATE DEVONIAN

Dipterus

MID–LATE DEVONIAN

Chirodipterus

EARLY DEVONIAN

orbit small

cheek long

Dipnorhynchus

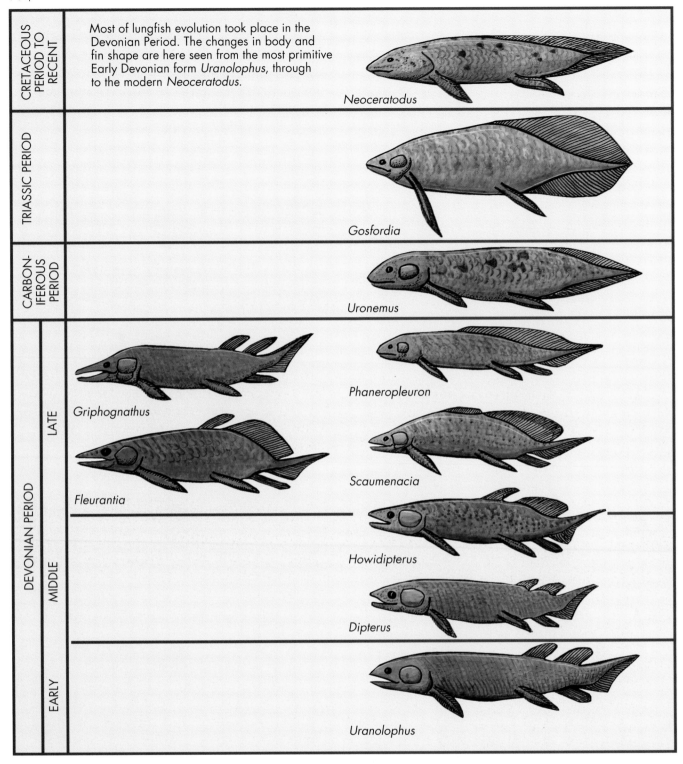

CRETACEOUS PERIOD TO RECENT	Most of lungfish evolution took place in the Devonian Period. The changes in body and fin shape are here seen from the most primitive Early Devonian form *Uranolophus,* through to the modern *Neoceratodus.*		*Neoceratodus*
TRIASSIC PERIOD			*Gosfordia*
CARBON-IFEROUS PERIOD			*Uronemus*
DEVONIAN PERIOD — **LATE**	*Griphognathus* / *Fleurantia*		*Phaneropleuron* / *Scaumenacia* / *Howidipterus*
MIDDLE			*Dipterus*
EARLY			*Uranolophus*

unchanged, as fossils of the Queensland lungfish from Lightning Ridge, New South Wales, show that the same species lived in Australia some 100 million years ago. Our story is largely concerned with that initial burst of evolutionary radiation back in the Devonian, when most of the radical changes took place.

▶ This limestone nodule found at Gogo by the author in 1986 has a complete fossil lungfish inside. After preparation the perfectly preserved three-dimensional skull of the long-snouted lungfish *Griphognathus whitei* was revealed.

LUNGFISH ORIGINS

From their first appearance the lungfishes are a highly characteristic group, easily recognised by a number of features in their skull roof and dentition. However, because of this their relationships with other bony fishes have been much disputed, although consensus now favours the hypothesis that they are most closely allied to the other fleshy-finned fishes, the Crossopterygii. Both early crossopterygians and dipnoans have cosmine and thick rhombic scales, and their fins have a well-developed internal skeleton. A possible link between these two groups is an enigmatic fish from the Early Devonian of Yunnan, China, called *Diabolepis*, meaning "devil scale" (first named as "*Diabolichthys*" in 1984). This fish has a skull roof pattern with a median "B" bone in front of the contacting "I" bones (as in crossopterygians), and the two sides of the lower jaws meet in an extensive, strong symphysis. The anterior external nostril is situated on the upper mouth margin as in dipnoans, and broad crushing tooth-plates are present. Despite the on-going debate over the relationships of *Diabolepis*, most now agree that it is undoubtedly a close link between the first true dipnoans and the crossopterygian fishes. Although the oldest true dipnoan, *Uranolophus*, comes from North America, the Early Devonian limestone rocks of southeastern Australia hold the greatest diversity of primitive dipnoans and possibly the key to understanding the early evolution of the group.

▲ Reconstruction of *Griphognathus whitei*, a long-snouted lungfish that broke down its food by rasping the large ventral gill-arch bones against a palate covered by many minute denticles.

▼ A superbly-preserved skull of the long-snouted lungfish *Griphognathus whitei*, from the Late Devonian Gogo Formation, Western Australia. Its long snout may have been used for nuzzling around the muddy sea-floor while using its electrosensory system to detect invertebrates.
Inset: palate view.

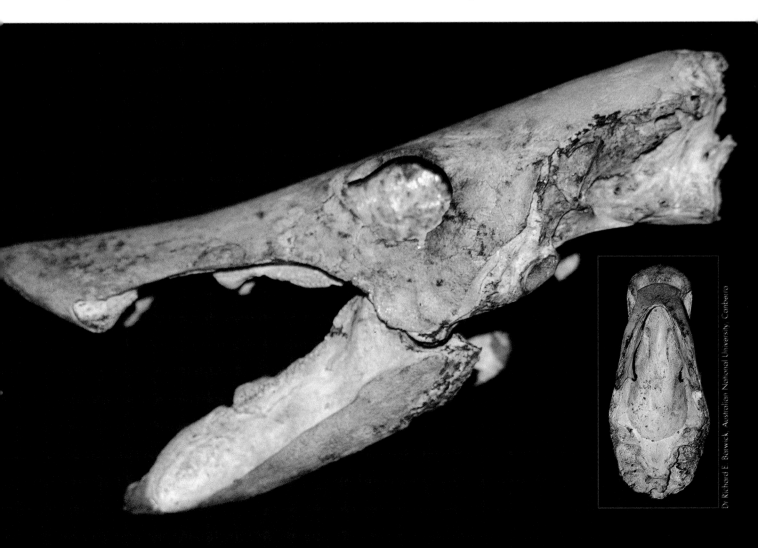

BASIC STRUCTURE OF PRIMITIVE DIPNOANS

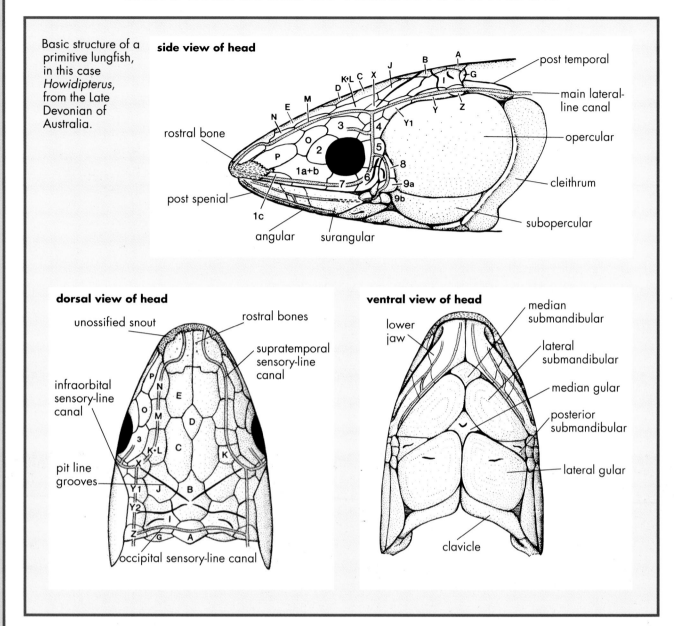

Basic structure of a primitive lungfish, in this case *Howidipterus*, from the Late Devonian of Australia.

side view of head

post temporal
main lateral-line canal
opercular
cleithrum
subopercular

rostral bone
post spenial
1c
angular
surangular

dorsal view of head

unossified snout
rostral bones
supratemporal sensory-line canal
infraorbital sensory-line canal
pit line grooves
occipital sensory-line canal

ventral view of head

lower jaw
median submandibular
lateral submandibular
median gular
posterior submandibular
lateral gular
clavicle

The one adaptation that seems to unite all the special anatomical features of early dipnoans is the ability to produce a powerful bite. The numerous tightly interconnected bones of the skull roof relate to powerful jaw muscle insertions on the inside of the skull roof. The massive size of the region where the lower jaws meet (the symphysis), the fusion of the palate to the braincase, the heavily built-up gill arch bones, and the specialised dental tissues, all reflect a skull capable of exerting great power in the bite.

The braincase of dipnoans is heavily ossified as a single piece, with the palate firmly fused to its lower surface. In primitive dipnoans the braincase has struts supporting the skull roof, creating large chambers for the passage of jaw muscles, which attach on the inside of the skull roof and run down to the lower jaw. The parasphenoid bone, which sits in the middle of the palate, is short and ploughshare shaped in primitive forms, although in later lineages it became expanded with a long posterior stalk. This elongation gave more room in the mouth for the gulping of air bubbles.

The snout of lungfishes has visible grooves for the incurrent nostril situated along the upper border of the mouth. The excurrent nostril opens from the nasal capsule directly into the palate, without any bones covering the nasal capsule.

Primitive dipnoans have a complex system of minute tubules running through the bone of the snout and ends of the lower jaws. Some scientists have interpreted these as an electrosensory system, like the detecting devices used by some modern fishes for finding food in muddy environments. Other scientists believe that the complex network of tubules in the snout were part of a nutritive system that fed the skin and sensory-line canals of the snout.

Lungfishes fed by one of two methods. They either had hardened tooth-plates for crushing food or a mouth covered with small denticles which were periodically shed (henceforth called "denticle shedders" or "denticulates"). The oldest dipnoans are characterised by having a powerful crushing bite, as seen in the large area where the lower jaws meet in the midline, and the large attachment areas for the jaw musculature. Primitive biters include forms with palates covered by shiny dentine, a tissue found below the enamel layer in most vertebrate teeth. These dentine-plated forms gave rise to tooth-plates with rows of teeth organised on tooth ridges. The denticle shedders, however, have powerful gill arch bones lined with smaller denticle-covered bones for rasping food against the denticle-covered palates. The gill arches of lungfish feature large ceratohyal elements and the hyomandibular does not take part in the jaw articulation, as in other osteichthyans. Many toothed bones accompany the gill arch series in denticle-shedding lungfishes.

The bodies of early lungfishes have two equally-sized dorsal fins, separate anal fin and heterocercal caudal fin, with long feathery paired pectoral and pelvic fins. Throughout the Devonian the trend was to change to having a shorter first dorsal fin with a longer second dorsal fin, with eventual merging of the median fins and tail fin. By the end of the Devonian, 355 million years ago, lungfishes had

The massive skull of *Dipnorhynchus sussmilchi*, one of the earliest known lungfishes, and certainly one of the best preserved species for its age anywhere in the world. This specimen was found embedded in shallow marine limestones outcropping at Burrinjuck Dam, near Taemas, in New South Wales, Australia.

John A. Long

acquired the body shape and fin plan that they were to keep for the rest of their evolution.

The soft anatomy of lungfishes is characterised by a number of unique features including specialisations of the nervous system such as concentrically layered olfactory bulb, and Mauthner cells in the brain. Lungfishes have a three-chambered heart, although partitioning in the atrium is only partial, and in *Neoceratodus* it is barely evident. Their lungs are developed as an outpocketing of the gut, developed from the primitive osteichthyan swim-bladder. The modification of the swim-bladder to form a functional lung was simply a matter of increasing the internal surface area for improved gas-exchange ability, and was thus not a complex evolutionary step.

The Early–Middle Devonian lungfishes had thick rhombic scales covered with cosmine, a shiny enameloid layer over the bones and scales, which housed a system of pores and interconnecting canals in the dentine beneath. The cosmine layer was subsequently lost in most dipnoans by the Late Devonian, and the bodies of these advanced fishes were covered by thinner, rounded scales. The snout of primitive cosmine-covered dipnoans was ossified as a stout single unit, which often broke away from the skull after death and can occasionally be found as isolated fossils (this is termed the "loose-nose problem" by Professor Erik Jarvik). As cosmine disappeared from the skeleton, the ossified snout was replaced by one formed of soft tissue, sometimes with special small bones covering its top surface.

DIPNOAN DIVERSITY: THE DENTICLE SHEDDERS

Dipnoans can be divided into two major groups, based on their methods of feeding: those having tooth-plates, and those having denticle-covered palates and lower jaws. The latter group were confined to the Palaeozoic Era and were most abundant during the Devonian Period. Tooth-plated dipnoans include those with true tooth-plates which grow by addition of new dental tissues from the margins of the tooth-plate and have a pulp chamber below the tooth-plate; and those with "primitive" dental plates and dentine-covered palates, often with simple, rudimentary tooth rows. These forms clearly represent primitive grades of organisation in the evolution of true dipnoan tooth-plates.

True dipnoan tooth-plates may comprise a great diversity of tissue types, many of which have been lost in later dipnoans. Heavily mineralised tissues such as "petrodentine" gave certain dipnoans the ability to develop extremely powerful crushing bites, able to crack and grind up hard-shelled prey items, such as primitive clams, lamp-shells and corals.

The earliest well-preserved lungfish is *Uranolophus*, from North America, an Early Devonian form with a denticle-covered palate. Its primitive nature is seen by the two equally-sized dorsal fins, thick cosmine-covered rhombic scales and a skull roof pattern with large "I" bones meeting each other behind the "B" bone, and numerous small bones forming the snout. The lower jaws meet in a strong contact zone, suggesting that *Uranolophus* had powerful jaw muscles for exerting much pressure when it bit.

In the Middle and Late Devonian, denticulate lungfishes existed side by side with tooth-plated forms, often together in the same environments; many of the fossil fish faunas of this age contain examples of both groups, indicating that the feeding strategies of the two groups did not seem to overlap. These denticulate dipnoans include the long-snouted rhynchodipterid family (for example *Griphognathus* and *Soederberghia*) which inhabited both marine and freshwater environments. Australia has several good examples of these Devonian dipnoans: the duck-billed lungfish

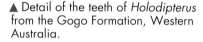

▲ Detail of the teeth of *Holodipterus* from the Gogo Formation, Western Australia.

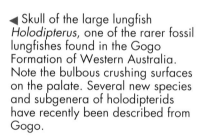

◄ Skull of the large lungfish *Holodipterus*, one of the rarer fossil lungfishes found in the Gogo Formation of Western Australia. Note the bulbous crushing surfaces on the palate. Several new species and subgenera of holodipterids have recently been described from Gogo.

John A. Long

Griphognathus whitei, a common fish at Gogo, in Western Australia; and the holodontid lungfishes, also known from Gogo, which include several species of the short-snouted, massively built *Holodipterus* as well as new forms currently being described by Professor Ken Campbell and his team in Canberra. *Soederberghia*, first described from the Late Devonian of East Greenland, is known in Australia from one skull roof found in red mudstones from a quarry at Jemalong Gap, near Forbes in New South Wales.

Griphognathus is an unusual-looking fish with a long, flat duck-like bill. Its strong gill arch skeleton has many muscle attachment scars indicating strong muscles to move the ventral gill arch bone (basibranchial) sideways and up and down, exactly like a file rasping away at a hard surface. This technique, coupled with the plier-like duck bill, may have enabled *Griphognathus* to snap off long pieces of branching coral

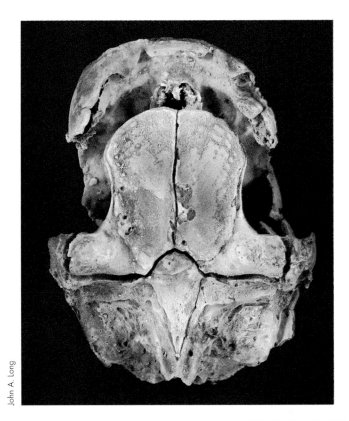

John A. Long

▲ Skull of the chirodipterid *Pillararhynchus longi,* showing unusually long dental plates that have rows of pointed teeth.

▶ The palate of *Chirodipterus australis,* showing dental plates used for crushing hard-shelled food items.

▼ These greyish-blue limestones outcropping along the Burrinjuck Dam have yielded some of the world's finest early lungfish fossils. The bones are freed from the rock using weak acetic acid to show their three-dimensional structures. Professor Ken Campbell, in the foreground, points to where he found the first complete skull of a *Dipnorhynchus.*

John A. Long

and then grind them up with the denticle-covered gill arch bones, palate and lower jaws. Alternatively, it may have used its long snout for nuzzling along the muddy sea floor, sensing for soft-bodied worms and other creatures. The success of *Griphognathus* is seen by the fact that it is one of the most widespread of all dipnoans in the Late Devonian, being found in North America, Europe and Australia. I'll never forget the day in August 1986 when I found a complete *Griphognathus* on my first season working at Gogo. First I picked up a large block showing only the shiny cosmine tip of the snout emerging, and knew immediately that I had a complete undamaged skull inside the rock. Then, only a metre or so away from this, I found another three blocks, each showing a cross-section of the body in the round. Incredibly the blocks all fitted together to make a large sausage-shaped nodule (see photograph on page 164). Inside was one of the world's most complete Devonian lungfishes. The preparation of the body and fins is still in progress.

The last known denticle-shedding lungfish is *Conchopoma,* from the Permian of Germany and North America, a specialised form with a tail fin like modern lungfishes and broad median palate bone (parasphenoid) covered in numerous denticles. In appearance *Conchopoma* resembles many of its contemporary tooth-plated lungfishes. The reason for these assumed parallel series of changes in differing lungfish lineages may be that both had a similar developmental plan. If juvenile fish of both groups follow the same pattern of growth and development (called ontogeny), then changes in the timing of this development can produce parallel changes in different groups. This form of evolutionary change is called heterochrony and is now used to explain how much of evolution was not necessarily driven by external environmental pressures but controlled by internal developmental factors within the organism. A clear evolutionary trend in some Devonian lungfishes is the retention of juvenile features into the adult stage of later species. This is called paedomorphosis and can explain why some features, such as large eyes and shorter cheeks, can evolve rapidly just by earlier sexual maturity within a lineage of fishes.

An alternative theory of lungfish evolution argues that the transition from denticulate palates to tooth-plates may not have been so great an evolutionary step. There are some denticulate lungfish that show massive dentine-covered palates with large tooth-like cusps (for example *Holodipterus gogoensis*), while other species found in the same site are slender-jawed with fine denticle shagreen covering the palate and biting areas of the lower jaw (*Holodipterus longi*).

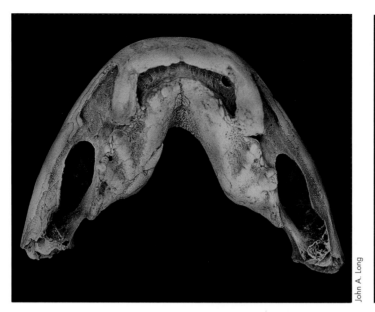

John A. Long

▲ *Speonesydrion iani*, an Early Devonian lungfish found near Taemas, New South Wales, Australia, represents the first stage in lungfish acquiring tooth-plates. The lower jaw shown here has developed rudimentary tooth rows.
▶▲ The palate of *Dipnorhynchus sussmilchi* shows the heavy layers of dentine and roughened tuberosities used to crush hard-shelled marine organisms.
▼ *Ichnomylax kurnai* is one of the oldest lungfishes known. It is represented solely by this damaged half of a lower jaw found in limestones exposed on the beach near Walkerville, Victoria, Australia.

DENTINE-COVERED PALATES: THE EARLY CRUSHERS

Three of the world's few known Early Devonian lungfish have been found in southeastern Australia. All of these are primitive lungfishes with powerful crushing dentitions formed of thick dentine sheets covering the palate and biting surface of the lower jaws. The skulls are known in two of these genera and show the primitive condition of the bone pattern, in having many small bones around the snout and front of the skull roof. The best known of these, *Dipnorhynchus* (meaning "two lungs snout"), is represented by three species, known from the Taemas–Wee Jasper and Cooma regions of New South Wales, and the Buchan district of eastern Victoria. All three species feature a heavily built palate with bulbous tuberosities used for exerting great pressure on the food, functioning much like a nutcracker to smash clams and other hard-shelled food items.

Speonesydrion (Greek meaning "Cave Island") is named after the site where the fossil was found, on Cave Island in Burrinjuck Dam, near Taemas. This fish has a skull roof pattern similar to *Dipnorhynchus*, but the palate and lower jaws have rudimentary rows of tuberosities, forming the primitive plan for a crushing tooth-plate. Similarly, *Ichnomylax,* known only from one side of the lower jaw found near Bell's Point (Waratah Bay), Victoria, has primitive tooth ridges and a bulbous dentine-covered crushing heel on the inside of the lower jaw.

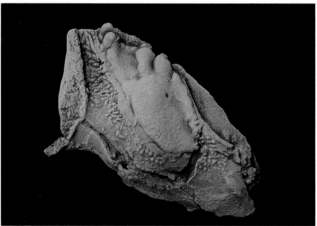

John A. Long

REFINED FEEDING MECHANISMS

Anyone who has watched a Queensland lungfish feeding will never forget the experience. Food is taken into the mouth, chewed up using the tooth-plates, and extruded out again as a pulpy long tube. Then it is chewed again and again, until all the material has been reduced to a readily digestible mass. The ability to feed in this manner must have been in use back in the Devonian when the first true tooth-plated dipnoans evolved. One group, the chirodipterid family, has tooth-plates that lack true teeth and have most of the tooth-plate lacking any cusps. These are termed "dental plates", as opposed to true "tooth-plates" that possess teeth (individual cusps, which are added on at the margins of the plates with continued growth). In Australia the chirodipterids are well represented by three species, all from the Late Devonian Gogo Formation of Western Australia.

Chirodipterus, first described from material in Europe and North America, is a commonly found genus in the Gogo Formation. *Chirodipterus australis* has broad crushing dental plates with weak tooth ridges, whereas *Gogodipterus paddyensis* has strongly developed tooth-ridges with deep grooves between them on each dental plate. The third Gogo species, *Pillararhynchus longi*, has a deeper skull than either of the other forms and possesses long, narrow tooth-plates with concave crushing surfaces. The palate bone (parasphenoid) has a patch of dentine on its front surface, indicating that it

took part in the crushing of food. The largest of the chirodipterids was a monster called *Palaedaphus insignis* from the Late Devonian marine deposits of Belgium. Its huge tooth-plates measure almost 14 cm long and 10 cm wide, suggesting that the fish may have grown to about 2 m in length.

Most of the known families of lungfishes have tooth-plates. The Devonian "dipterids" represent the most primitive

▲ Restoration of *Dipnorhynchus sussmilchi*, Australia's earliest fossil lungfish. Based on actual fossil skulls and scales with body shape restored from information based on other primitive Early Devonian lungfishes.

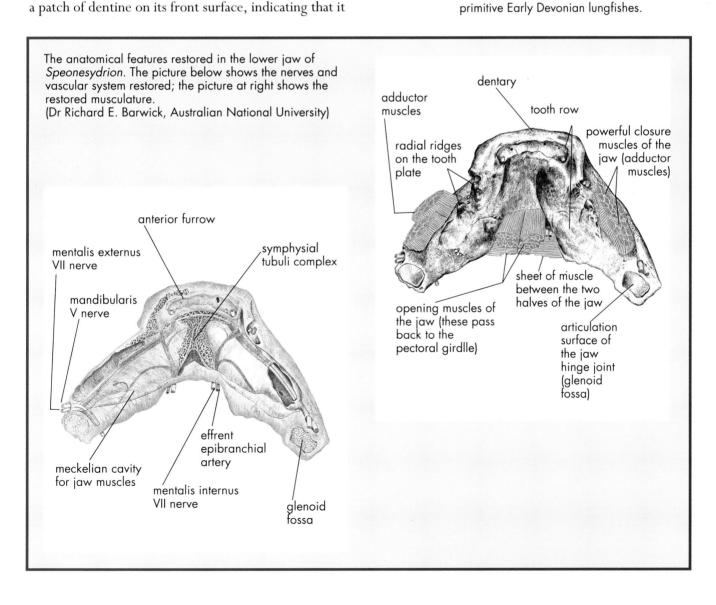

The anatomical features restored in the lower jaw of *Speonesydrion*. The picture below shows the nerves and vascular system restored; the picture at right shows the restored musculature.
(Dr Richard E. Barwick, Australian National University)

adductor muscles

radial ridges on the tooth plate

dentary

tooth row

powerful closure muscles of the jaw (adductor muscles)

opening muscles of the jaw (these pass back to the pectoral girdlle)

sheet of muscle between the two halves of the jaw

articulation surface of the jaw hinge joint (glenoid fossa)

anterior furrow

mentalis externus VII nerve

mandibularis V nerve

symphysial tubuli complex

meckelian cavity for jaw muscles

effrent epibranchial artery

mentalis internus VII nerve

glenoid fossa

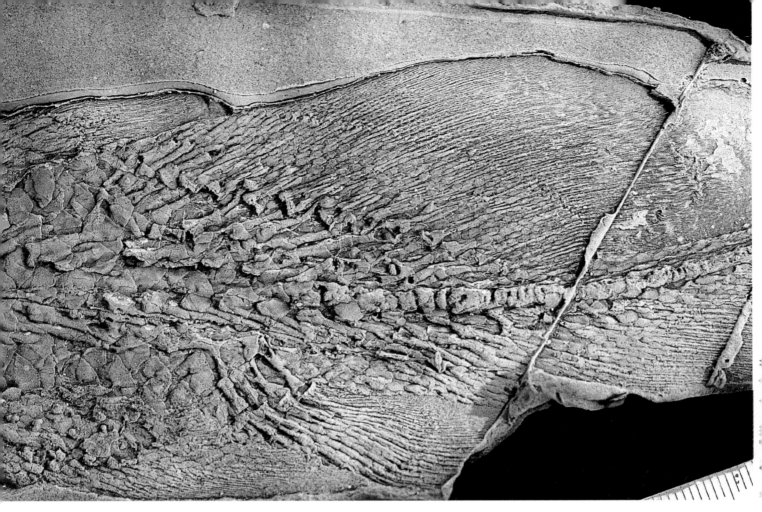

▲ Detail of the tail fin of *Barwickia downunda*, showing part of the well-ossified axial skeleton.

▼ *Howidipterus*, a Late Devonian lungfish from Mount Howitt, Victoria, Australia. This specimen is a latex peel taken from the prepared mould of the fish skeleton preserved in shale.

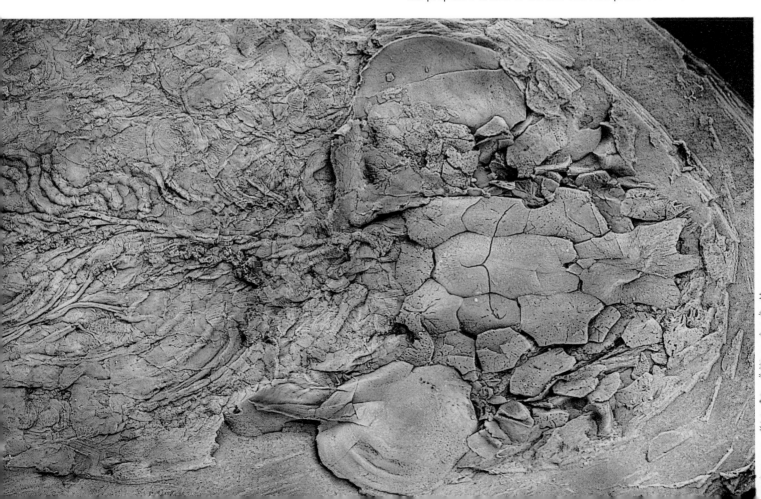

THE ORIGIN OF LUNGS

The earliest fishes were devoid of internal hydrostatic organs such as the swim-bladder. We know this from studying the primitive jawless fishes alive today (the hagfish and lamprey) and fossil jawless fishes, most of which were armoured forms that probably lived near the bottom of their lake, river or sea. The origins of the swim-bladder, an internal organ that enables a fish to regulate its vertical position in the water column, is associated with a major evolutionary radiation for the fishes—the origin of the osteichthyans, or true bony fishes. This was accompanied by radical changes in their shape, as they no longer relied on wide pectoral fins to gain lift in the water column. Instead the fins begin to serve different purposes, mainly to improve manoeuvrability and braking. Thus from the shapes of fossil fishes and comparison with their living relatives, the osteichthyans dating back to the Late Silurian mark the first occurrence of swim-bladders in fishes.

How the swim-bladder became a lung is really a simple morphological step—the organ already existed as a means of gas regulation—and by complex folding of the internal surface of the organ and intense vascularisation, or numerous blood capillaries, such surfaces became excellent respiratory surfaces. In most osteichthyans the blood flowing to the swim-bladder had already been oxygenated, as the arteries supplying it came from the posterior efferent branchial arteries or directly from the dorsal aorta. In fishes that live in oxygen-depleted environments the ability for the swim-bladder to take in oxygen is quite advantageous. In lungfish such as *Protopterus* the blood flowing to the swim-bladder (now called a lung) comes from the ventral and dorsal aortae and the pulmonary arteries, and these supply the organ with only partially oxygenated blood. To take air into the swim-bladder and allow the blood to absorb oxygen also requires some advanced development of

How lungs evolved from osteichthyan swim-bladders. Lateral views above, cross-sections below.

The primitive osteichthyan, a ray-finned fish. The swim-bladder is an outpocketing from the gut.

A lungfish, *Neoceratodus*

A tetrapod with paired lungs

musculature to squeeze the air inside them and force it under pressure into the lung. In most cases such fishes simply swallow air, which goes into the lung via a pneumatic duct.

Lungs were the next progressive step in the modification of the swim-bladder as it became more efficient at gaseous exchange. Some modern bony fishes, such as the mudskippers, can utilise their swim-bladders to take in oxygen directly from the air for short periods, and other fish use the swim-bladder for sound production or sound reception. The major step in going from a swim-bladder to a respiratory organ probably occurred at least twice in osteichthyan evolution—once for the group of crossopterygian fishes leading to early amphibians (the Osteolepiformes), and once within lungfishes. Evidence for this is seen in the presence of cranial ribs in lungfish fossils, which suggest the timing of lung development and air-gulping as lungfishes moved away from the seas and into freshwater habitats. All primitive marine lungfish appear to lack cranial ribs. In modern lungfish the cranial ribs anchor the pectoral girdle during the act of air-gulping, so the origin of cranial ribs in fossil lungfishes can be tied in to the beginning of air-gulping in fossil species. The ability for some lungfish to aestivate (stay in burrows and await the next rainy season) dates back to the Permian, based on fossil lungfish burrows, sometimes with the fish still inside.

The first record of crossopterygian fishes being able to breathe air is uncertain and can only be based on the assumption that the presence of an internal palatal nostril (choana) in tetrapods and fishes denotes the ability to breathe air. If this is correct then the osteolepiform fishes began "occasional" air breathing at the start of the Middle Devonian, the same time that lungfishes took to air-gulping. Both lungfishes and crossopterygians apparently began air-breathing when they moved from marine to freshwater habitats.

▲ *Dipterus*, a Middle Devonian lungfish from Scotland, is one of the oldest lungfishes known to have cranial ribs. This suggests that such fishes were capable of air-gulping. *Dipterus* was also the first fossil lungfish to be scientifically described, by Sedgwick and Murchison in 1828.

▶ *Scaumenacia*, a Late Devonian tooth-plated lungfish from Miguasha, has resemblances to the Australian *Howidipterus* but is more advanced in having an elongated first dorsal fin and in several features of the skull.

▼ A 370 million-year-old lungfish, *Chirodipterus australis*, from the ancient reef deposits of Gogo, Western Australia. Lungfish started off as marine gill-respiring fishes and later shifted to freshwater habitats, evolving the ability to breathe air as an aid to gill respiration. Ultimately some species acquired the ability to aestivate in burrows, away from the water, while waiting for seasonal rains to come.
John A. Long

John A. Long/Dr Marius Arsenault, Parc de Miguashua

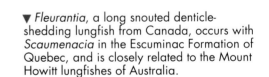

▼ *Fleurantia*, a long snouted denticle-shedding lungfish from Canada, occurs with *Scaumenacia* in the Escuminac Formation of Quebec, and is closely related to the Mount Howitt lungfishes of Australia.

grade, as these fishes have two dorsal fins and possess shiny cosmine on the dermal bones. The "fleurantiids" and "phaneropleurids" are more advanced than the dipterids, in that they have the first dorsal fin much reduced and the second dorsal fin enlarged, and they possess simplified skull roof and cheek patterns. Within the general evolution of Devonian lungfishes we see the reduction of dorsal fins and merging of the anal fin with the caudal fin, to give the same appearance as the modern Queensland lungfish. Steps in achieving this plan are seen in the transformation series going from *Uranolophus* to *Dipterus* to *Howidipterus* to *Scaumenacia* to *Phaneropleuron (page 164)*.

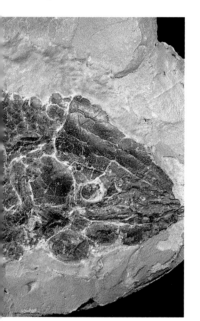

▶ Reconstruction of *Barwickia downunda*, a Late Devonian lungfish from Mount Howitt, Victoria, Australia.

▼ The skull of *Howidipterus* shows much variation in the shapes of bones and the degree of fusion between bones. This is a latex cast of the cleaned fossil that has been whitened with ammonium chloride to highlight the surface details.

The final successful body plan was achieved by the end of the Devonian, as seen in the Scottish genus *Phaneropleuron* and retained in all later lungfish.

Howidipterus, from the Late Devonian of Victoria, has unusual tooth-plates with well-developed teeth along the margins of each plate and smooth crushing surfaces towards the centre of the plates. The skull has many primitive features such as the retention of the large "D" and "K" bones. The parasphenoid has a well-developed stalk, unlike the primitive small diamond-shaped type seen in *Dipterus*. A recent study of two lungfishes from the Late Devonian Mount Howitt site, Victoria, shows that although the two have superficially differing dentitions—*Howidipterus* has tooth-plates, whereas

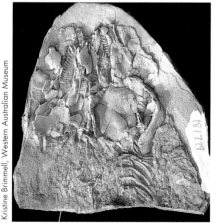

Kristine Brimmell, Western Australian Museum

Barwickia appears to be a denticle-shedder—both have identical body shapes, similar numbers of ribs and a similar plan of fin support bones. And when the dentitions are studied more closely, it appears that they have similar types of tooth-plates, one being dominated by rows of teeth with few denticles (*Howidipterus*), the other having few rows of teeth but many denticles (*Barwickia*). They are now regarded as both members of the fleurantiid group. The dentitions of other fleurantiids, such as *Andreyevichthys* from Russia, show that both kinds of dentition can exist in the same species as a matter of growth variations. Such precise similarities in the postcranial skeletons of the two Mount Howitt forms are unknown in other fossil lungfishes and strongly suggest that they evolved from a common ancestor with a similar body plan, into two distinct forms having different feeding strategies. The Mount Howitt lungfishes lived in a large lake environment.

Modern fish communities living in such lake environments are often based on a common ancestor entering the lake and then speciating into many similar forms, each with a slightly different feeding strategy (the lake-dwelling cichlids of Africa are an example).

All dipnoans of the Carboniferous Period had only one continuous median fin that merged with the tail fin, while their tooth-plates may be quite specialised with numerous tooth rows and closely packed cusps (for example *Ctenodus*). The only Carboniferous dipnoan described from Australia is *Delatitia breviceps,* first named as a new species of the European genus *Ctenodus* by Arthur Smith-Woodward in 1906. Recent preparation of the original specimen, from Mansfield, Victoria, has revealed that the sensory-line canals differ in the Australian form, as do other features of the skull roof pattern, and so in 1985 it was redescribed and assigned to a new genus. Well-preserved lungfishes of the Carboniferous and Permian Periods of Europe and North America include *Conchopoma*, a denticulate form, and *Uronemus*, with modified narrow tooth-plates. These and all other known subsequent lungfishes had achieved the body and fin pattern that still exists in modern lungfishes. The North American Permian form *Gnathorhiza* is famous for being found preserved within its fossil burrows, testifying to the ability of some lungfishes of Palaeozoic times to aestivate.

By the Mesozoic Era the majority of lungfishes were ceratodontids (the group that includes our living lungfish, *Neoceratodus*) or lepidosirenids (the group including the living African and South American lungfishes).

One of Australia's best-preserved Triassic lungfishes is *Gosfordia truncata,* from the Hawkesbury Sandstone near Gosford, New South Wales. It was first described and named by British palaeontologist Arthur Smith-Woodward from about five incomplete specimens, and later redescribed by Alex Ritchie of the Australian

◀◀◀ *Gosfordia truncata,* a complete fossil lungfish found in Triassic rocks near Gosford, New South Wales. *Gosfordia* is a close relative of the lungfish lineage leading to *Neoceratodus forsteri,* the modern Queensland lungfish.

Museum from a superb new specimen found in 1980 by a quarryman, Mr John Costigan. *Gosfordia* has a deep, plump body and broad tail, and in overall length was about 50 cm. The body of *Gosfordia* suggests that it was a strong swimmer, and not adapted for aestivation like the South American and African lungfishes.

Most other Mesozoic lungfishes from Australia are

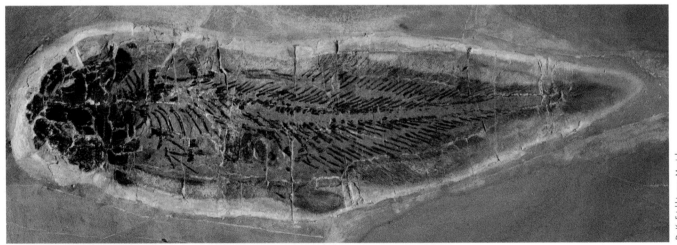

Dr. K. Frickhinger, Munich

John A. Long

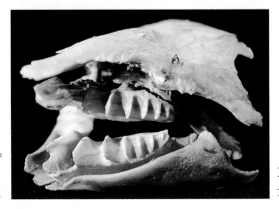

John A. Long

▲ *Conchopoma gadiforme,* a Permian lungfish found in Europe and North America. This specimen comes from Pfalz, Germany.

◀ Skull of the living Queensland lungfish, *Neoceratodus forsteri.*

◀◀ A fossil tooth-plate of the lungfish *Arganodus tiguidensis* from the Early Cretaceous of Niger Republic.

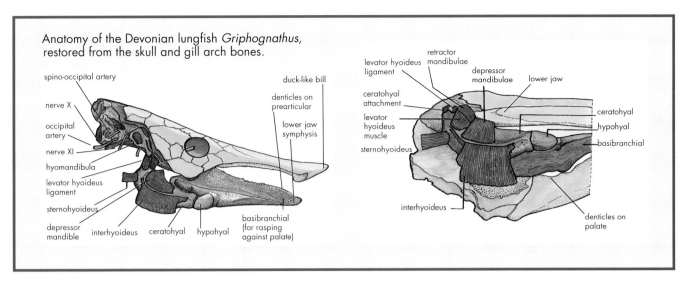

Anatomy of the Devonian lungfish *Griphognathus*, restored from the skull and gill arch bones.

represented by only isolated tooth-plates, such as the *Ceratodus* tooth-plates found in the Triassic Blina Shale of northern Western Australia, the Rewan Formation (of southern Queensland) and the Wianamatta Group (of New South Wales). Tooth-plates found in Early Cretaceous rocks indicate that the extant Queensland lungfish, *Neoceratodus forsteri,* was then living near Lightning Ridge, while in Victoria *Ceratodus nargun* survived in rivers of the cold rift valley that formed as Antarctica and Australia were beginning their separation.

During the Mesozoic Era, lungfish skull roof bones become simpler and there is great variation in tooth-plate morphology, even within an individual genus such as *Neoceratodus.* In the Tertiary, forms such as *Neoceratodus gregoryi* may have reached lengths in excess of 3 m, based on large tooth-plates and skull roof bones found from the Miocene river and lake deposits of central South Australia—and at least four other *Neoceratodus* species were widely distributed in the lakes and rivers of central Australia and Queensland. Although many species are known only from tooth-plates, the Redbank Plains area of southern Queensland has yielded semi-articulated skull and

body remains of *Neoceratodus denticulatus*, which are probably Eocene in age. These were first described in 1941 by Professor Edwin Sherbon Hills of the University of Melbourne.

Today the Dipnoi are known from three surviving genera: *Proptopterus* in Africa, *Lepidosiren* in South America and *Neoceratodus* in Queensland, Australia. Of these the Queensland lungfish is by far the most primitive; it has changed little, if in fact at all, in more than 100 million years. The lepidosirenids are longer, more slender fish with greatly reduced pectoral and pelvic fins for sensory functioning. During the dry season *Protopterus* can burrow into the ground and await the next rainy season, and is therefore adapted to survive in harsh climatic conditions where the Queensland lungfish would not survive. However, despite this, the Queensland lungfish has amazingly survived Australia's harsh climatic changes and is still Australia's only primary freshwater fish—a species that has evolved here rather than being an immigrant. We can only hope that its future for the next 100 million years is not endangered by humankind or our pollution of the planet.

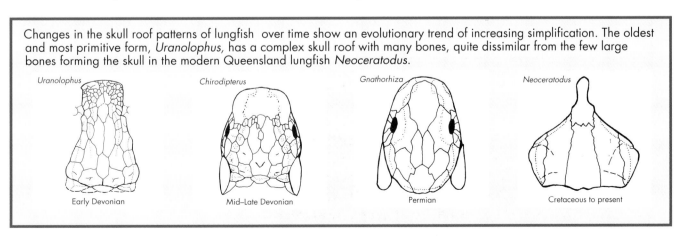

Changes in the skull roof patterns of lungfish over time show an evolutionary trend of increasing simplification. The oldest and most primitive form, *Uranolophus*, has a complex skull roof with many bones, quite dissimilar from the few large bones forming the skull in the modern Queensland lungfish *Neoceratodus*.

Chapter 9

BIG
TEETH,
STRONG
FINS

SUBCLASS CROSSOPTERYGII

The lobe-finned fishes, Crossopterygii, were a major group of large predators back in the Devonian Period but are today represented by a sole surviving species, the coelacanth *Latimeria chalumnae*. Several types of crossopterygians arose by the Middle Devonian, including the osteolepiforms, the group from which the first four-legged animals would evolve. All the crossopterygians were highly active predators, some of which had evolved highly refined dental tissues like enamel on their large stabbing teeth. Although some crossopterygians such as porolepiforms and onychodontiforms achieved a peak of diversity during the Devonian and then mysteriously disappeared, others reached their acme in the Carboniferous (coelacanths and rhizodontiforms). Some of these rhizodontiform fishes reached estimated sizes up to 6 m in length and dominated the lakes and rivers of their day. The only crossopterygians to survive the Palaeozoic Era are the coelacanths, which quickly reached their modern form by the start of the Mesozoic Era, and have remained relatively unchanged ever since.

J. Schaur/ Hans Fricke, Max Planck Institute

Previous page: Laccognathus, a well-preserved Late Devonian holoptychioid porolepiform from the Lode site in Latvia. This specimen is shown preserved with the mouth open. *Dr Oleg Lebedev, Palaeontological Institute, Moscow.* Inset: *Koharolepis*, a Middle Devonian canowindrid osteolepiform from Mount Crean, southern Victoria Land, Antarctica. These fishes are thought to be the most primitive of all known osteolepiforms. *John A. Lang*

▲ The coelacanth *Latimeria chalumnae*, the only living crossopterygian fish, was discovered in 1938 off the coast of South Africa. This rare photo shows a living coelacanth in its natural habitat near the Comoro Islands. The coelacanths live at depths of between 100 and 300 metres and can operate their lobed pectoral and pelvic fins in independent movements, in exactly the same manner as a four-legged land animal.

BASIC STRUCTURE OF CROSSOPTERYGIANS

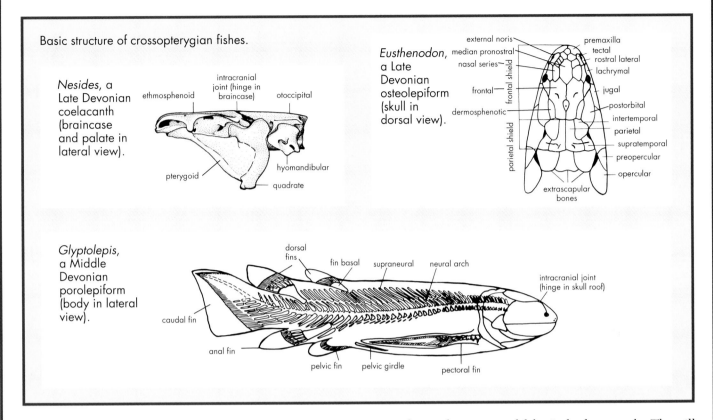

Basic structure of crossopterygian fishes.

Nesides, a Late Devonian coelacanth (braincase and palate in lateral view).

ethmosphenoid — intracranial joint (hinge in braincase) — otoccipital

pterygoid — quadrate — hyomandibular

Eusthenodon, a Late Devonian osteolepiform (skull in dorsal view).

external noris — median pronostral — nasal series — frontal — dermosphenotic — parietal shield — premaxilla — tectal — rostral lateral — lachrymal — frontal shield — jugal — postorbital — intertemporal — parietal — supratemporal — preopercular — opercular — extrascapular bones

Glyptolepis, a Middle Devonian porolepiform (body in lateral view).

dorsal fins — fin basal — supraneural — neural arch — intracranial joint (hinge in skull roof)

caudal fin — anal fin — pelvic fin — pelvic girdle — pectoral fin

Crossopterygian fishes are characterised by being relatively long-bodied with a skull that in primitive forms has a hinged braincase, divided into a front section (ethmosphenoid) and rear section (oticco-occipital). This flexure within the braincase is reflected in the skull roof bones of each group of crossopterygians; they are said to have a frontal and a parietal shield separated by the intracranial joint. Some advanced members of the groups have the intracranial joint immobilised by fusion of bones between the frontal and parietal shields (for example panderichthyids, discussed in Chapter 10). The head has large eyes in most species, and paired external nostrils in all groups except for osteolepiforms, which have a single external pair of nostrils and a palatal nostril opening called the choana.

The cheek has a regular pattern of bones, with one or more large squamosal bones present, and the jaws have large dentary bones supported by a series of infradentaries. There are well-developed fangs in addition to regular marginal teeth on the toothed bones —except in coelacanths, which have a smaller area of toothed biting jaws and an unusual double-tandem jaw joint (clearly a specialised feeding mechanism).The teeth have enamel present over dentine, and in some groups, like the osteolepiforms, porolepiforms and rhizodontiforms, there is complex infolding of the enamel and dentine. In cross-section these teeth appear highly complex and are termed labyrinthodont teeth. The gill arches are well ossified and have large ventral gill bones (basibranchials), sometimes with a forward-pointing sublingual bone.

The shoulder girdles are well ossified, as is the internal shoulder girdle (scapulocoracoid). All crossopterygians have well-ossified pectoral and pelvic fin skeletons, although this is most highly evolved in the osteolepiforms and rhizodontiforms, which have a powerful humerus that articulates with ulna and radius bones, as in all groups of higher vertebrates (except limbless forms). The bodies are often rather long in crossopterygians and rather conservative in shape; no unusual deep-bodied or flattened forms seemed to evolve.

The scales and dermal bones in all primitive crossopterygians have cosmine present, but this is often lost in later lineages. The scales are characteristic for most crossopterygian groups. Primitive members of the osteolepiforms and porolepiforms have thick, rhombic cosmine-covered scales, but during their separate evolutionary radiations these change to rounded non-cosmine-covered scales. The rhizodontiforms, onychodontiforms and actinistians have rounded thinner scales. The sensory-line systems of crossopterygians are well developed and can be seen as a series of pores and deep pit-line canals on the dermal bones and scales.

One of the most remarkable biological discoveries of the twentieth century took place in South Africa in the late 1930s. A strange-looking fish about 1.5 m long was caught off the mouth of the Chalumna River, near East London, South Africa, on 22 December 1938. The following day Marjorie Courtenay-Latimer, a young ichthyology curator at the East London Museum, saw the fish while searching for unusual specimens at the dockside. The fish was bizarre—it had muscular fleshy lobes for the pectoral, pelvic, second dorsal and anal fins. The tail was symmetrical with a long central lobe and a tuft at the end.

Once she saw that this was indeed an unusual fish, she then realised the problem in getting it back to her laboratory. The fish was already beginning to smell as decay had set in, but wrapping it in sack cloths, she

▲ *Caridosuctor* ("shrimp-eater") from the Early Carboniferous Bear Gulch Limestone of Montana, USA.

managed to persuade a taxi driver to take her and the prize fish back to the museum. Some days later, when Professor J.L.B. Smith, Senior Lecturer in Chemistry, inspected the fish, he was able to identify it as a member of a group called coelacanths, thought to be extinct for more than 50 million years! Smith named the fish *Latimeria chalumnae* in honour of Miss Latimer and the location the fish came from.

By this time the specimen had been gutted and little scientific information could be obtained from it, so the search began for another coelacanth. Not until 1952 did the next specimen come to light, this time from the Comoro Islands northwest of Madagascar, about 2000

Features of the skull of the primitive crossopterygian *Youngolepis*, from the Early Devonian of China. Left, in ventral view; right, in dorsal view.

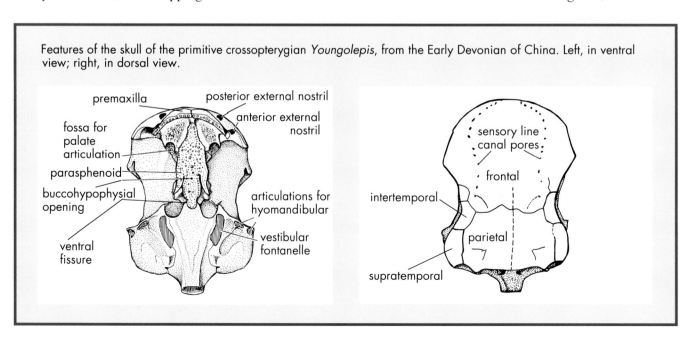

km north ofthe first specimen's location. Since then a good number of coelacanths have been caught, mostly around the Comoro Islands. The anatomy of the coelacanth was first described in detail by French professors Millot and Anthony, and in recent years many other scientists have studied detailed aspects of its biology. Perhaps the greatest advances in coelacanth studies have been made by Professor Hans Fricke of the Max Planck Institute, Germany, who filmed several living coelacanths in their natural habitat using a small submersible submarine, at depths of between 117 and 198 m. It had taken Fricke nearly 40 dives at 30 different sites to find the elusive coelacanth in its natural home. *Latimeria* is quite a slow-moving fish, which often drifts in the currents and adjusts its position using very mobile pectoral and pelvic fins. The fins can move like the gait of a land animal—pectorals and pelvic have the ability to move in opposite directions. Coelacanths live at depths between 100 and 700 m and feed on other fishes and the occasional cuttlefish.

The significance of the discovery made by Latimer and Smith in the late 1930s can really be appreciated when we consider what was known about fossil coelacanths. They were first recognised from fossils by Swiss scientist Louis Agassiz, and in his famous books of 1843–1844, *Recherches sur les poissons fossiles* he described *Coelacanthus granulatus* from the Late Permian of Germany. A large number of fossil coelacanths were

▼ The most primitive coelacanth that we know of is *Miguashaia bureaui,* from the Late Devonian Escuminac Formation of Quebec, Canada. It is one of two coelacanths that retain the primitive heterocercal tail, as all later forms have a "tufted tail" with an elongated central lobe.

then identified in rocks ranging from Middle Devonian age to the end of the Mesozoic Era, but the fact that no Cenozoic fossil coelacanths had been found suggested that the group became extinct along with the dinosaurs at the end of the Cretaceous. The coelacanth was first promoted as a "missing link" between fishes and land animals. In fact the link it does represent is as the only surviving member of a once large and diverse group of fishes called the crossopterygians ("tassle-finned fishes"). Within the Devonian members of this group are to be found the closest fish groups to the earliest land animals, the amphibians.

The crossopterygians include five major groups, of which only one (Order Actinistia, the "coelacanths") survives today, known from just one species. The extinct groups are the osteolepiforms, porolepiforms, onychodontiforms and rhizodontiforms. Aside from these well-defined groups, there are some "stem-group" crossopterygians that are intermediate in form between some of these groups and clearly represent the primitive character states for the group as a whole. These have been briefly mentioned in Chapter 6 and are discussed below as "primitive crossopterygians".

PRIMITIVE CROSSOPTERYGIANS

The oldest fossil crossopterygians date from the Early Devonian. By this time only the porolepiforms and some primitive crossopterygians of uncertain affinity had appeared. The first true members of all the other crossopterygians appeared shortly after in the Middle Devonian, indicating a rapid radiation of the group overall. The Early

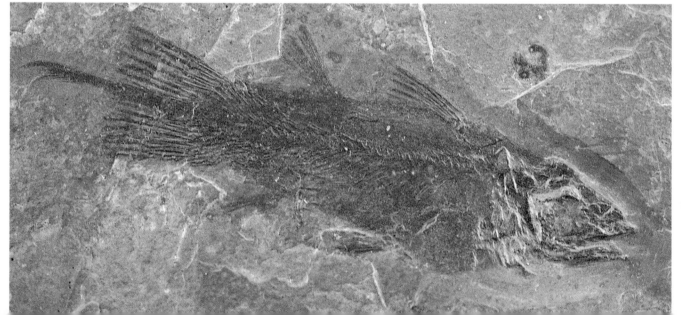

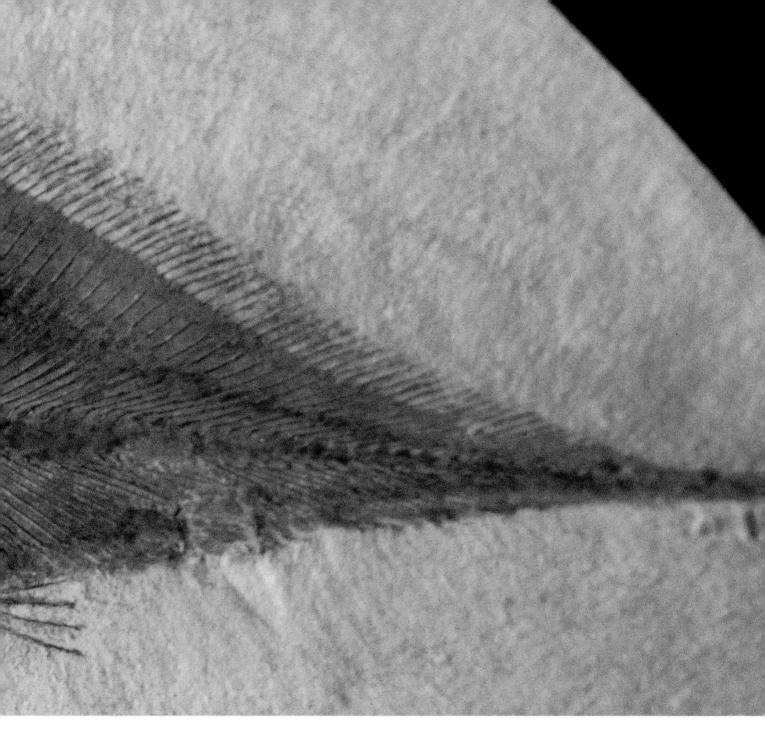

▲ The unusual coelacanth *Allenypterus* from the Bear Gulch Limestone of Montana, USA, was so bizarre that it was first described as being a ray-finned fish.

◄ *Osteopleurus newarki*, a coelacanth from the Jurassic of North America.

Devonian primitive crossopterygians include two similar forms called *Youngolepis* and *Powichthys*, already discussed in Chapter 6. *Youngolepis* is known from the Early Devonian of China and was named in honour of the famous Chinese palaeontologist C.C. Young, whereas *Powichthys* is named after its discovery site, Prince of Wales Island (initials POW), in Arctic Canada. They were small predatory fishes, no larger than about 30 cm maximum length, with heavily ossified bones in the head and thick rhombic scales over the body. They appear to be more advanced than other crossopterygians in having a series of many small bones flanking the main paired bones of the skull roof, a condition also found in lungfishes. To compensate for their small eyes they had a well-developed lateral-line sensory system to detect prey.

Both of these forms are known principally from their cosmine-covered skulls, which have a relatively long frontal shield and a short parietal shield. The braincase in both forms is not fully divided into two components, as in other crossopterygians, indicating a primitive condition that precedes the division of the braincase into two components in all other crossopterygians.

Youngolepis is better known than Powichthys and has a cheek showing several fused bones, but is basically much like that of an osteolepiform in its cheek bone pattern. Youngolepis and Powichthys have many small bones along the flanks of the large central skull roof bones. In addition, the flask-shaped cavities within canals within the cosmine layer have enameloid tissue dipping into them, a feature seen developed even further in porolepiforms, but not at all in other crossopterygians. Recent work focusing on the histology of the bones and teeth of Youngolepis and Powichthys suggest that they should be included as primitive members of the porolepiform group and that they are actually closer to lungfishes than to other crossopterygians.

COELACANTHS (ACTINISTIA): THE TASSEL-TAILS

The Actinistia, or coelacanths, first appear in the Middle Devonian and survive today, represented by the genus Latimeria. The peak of diversity for coelacanths was probably achieved during the Carboniferous, when many varied forms inhabited the shallow seas and rivers of the world. The most distinguishing features of actinistians is that they lack an upper jaw bone (maxilla), have a loose cheek bone arrangement, numerous paired snout bones each with large pores (reflecting the fact that the snout has a special rostral organ), and the lower jaw has a special "double-tandem" articulation. The bodies of coelacanths have a long median lobe on the tail with a rear tassel or tuft, and the paired fins and some of the median fins are all strongly lobed. Those fins not strongly lobed in some genera have stout spines supporting the fin web. The shoulder girdle bones are more elongated compared to other crossopterygians, and most genera possess an additional bone unique to coelacanths, the extracleithrum, attached to the base of the cleithrum.

The earliest fossil coelacanths come from the late Middle or early Late Devonian of Germany, Canada and Australia. Forms such as Diplocercides show the typical suite of coelacanth characteristics but lack many of the advanced coelacanth features, such as specialisations of the braincase and having an equal number of tail fin rays as supporting bones in the tail. Miguashaia, from the Late Devonian Escuminac Formation of Quebec, Canada, has many primitive coelacanth skull features and a heterocercal tail as in other crossopterygians. It is regarded as the most primitive of all coelacanths that are reasonably well known. An as-yet-undescribed Late Devonian coelacanth from Mount Howitt, Australia, is known only from a fragmentary skull and parts of the body and tail. It appears similar in many ways to Miguashaia.

The extraordinary coelacanths from the Early Carboniferous Bear Gulch Limestone of Montana studied by Dick and Wendy Lund show the extreme adaptations of body form that coelacanths achieved. Some slender forms, like Caridosuctor (meaning "prawn eater"), were similar to later coelacanths in their body form, whereas Hadronector, Lochmocercus and Polyosteorhynchus were fairly squat forms. The most extreme body shape is seen in Allenypterus, a deep-bodied coelacanth with a long tail region and rather small dorsal, pelvic and anal fins. Allenypterus is so unusual that it was first described as a ray-finned fish and not a coelacanth!

During the Mesozoic the coelacanths became fairly conservative, although some managed to reach large sizes, like Mawsonia from the Cretaceous of South America and Africa. The largest species, Mawsonia gigas, may have reached lengths in excess of 3 m, and was the one of the largest predatory lobe-finned fish in the shallow seas of Gondwana at this time.

Dr Marius Arsenault, Parc de Miguashua

▲ *Quebecius quebecius*, a porolepiform from the Late Devonian Escuminac Formation of Quebec, Canada.

▶ Reconstruction of *Quebecius quebecius*, a Late Devonian porolepiform fish known from Quebec, Canada (where else with a name like that?). (After the work of Dr Hans-Peter Schultze).

▼ The head of *Porolepis*, an Early Devonian cosmine-covered porolepiform.

POROLEPIFORMES: FAT-HEADED, BEADY-EYED PREDATORS

The Porolepiformes were relatively large predatory fishes that lived in the Devonian Period. They have been found at sites all around the world but are most prevalent in Middle and Late Devonian freshwater deposits. The group takes its name from having rows of pores on the cosmine-covered scales. Other features that characterise the porolepiform fishes are a broad skull with small eyes, the presence of a prespiracular bone in the cheek, and a special style of infolding of the enamel and dentine in the large fangs, called dendrodont tooth structure. At the front of the lower jaws is a large whorl of stabbing teeth. From the shape of their bodies and tail, most porolepiforms were thought to be ambush predators, lying in wait for passing prey which was caught by a quick forward lunge.

The oldest form is *Porolepis* from the Early Devonian of Spitzbergen and western Europe. *Porolepis* reached lengths of about 1.5 m, fairly large for its time. It had a thick cosmine cover on all its bones and scales, and the eyes were very small. *Porolepis* and a few other forms known from fragmentary remains, which had thick rhombic scales, are placed in the

◀ *Holoptychius*, a widespread large porolepiform that flourished towards the end of the Devonian Period. This specimen comes from Scaumenac Bay, Quebec, Canada.

▼ *Holoptychius* specimens from Dura Den, Scotland.

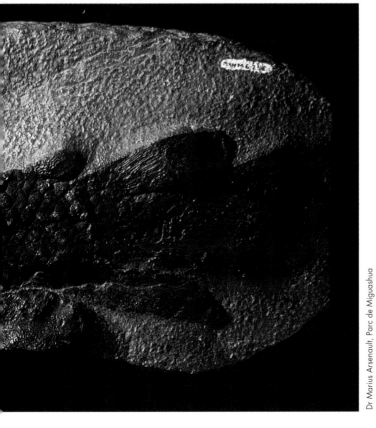

John A. Long

Dr Marius Arsenault, Parc de Miguashua

John A. Long

family Porolepidae, and most Middle and Late Devonian porolepiforms are placed in the family Holoptychiidae, as these lack cosmine and have rounded scales. *Porolepis* was among the largest predators of the Early Devonian seas and near-shore environments. Their sluggish appearance was no doubt efficient by comparison with the heavily armoured primitive placoderms and acanthodians, upon which they most likely preyed.

In the Middle Devonian the chief group of porolepiforms emerged, the holoptychioids. These fishes grew to enormous sizes for their day, maybe 2.5 m or more, and were fearsome ambush predators. Although some occur in marginal marine deposits, most had invaded the river systems, away

▲ Skull of *Onychodus*, a predatory dagger-toothed fish from the Late Devonian Gogo Formation, Western Australia.

from the giant predatory dinichthyid placoderms, where they could be the top predators in the water. The holoptychioids are considered more advanced in evolutionary terms than the porolepids, in that they have lost the thick cosmine cover on the bones and scales, the scales have become rounded, the skull has a specialised set of bones around the external nostrils (including a "nariodal" bone), and the mouth bears enlarged tooth whorls at the front of the lower jaws. These tooth whorls differ from those of the onychodontids because they have a series of large

▶ Reconstruction of *Holoptychius*, a large freshwater predator of the Late Devonian.

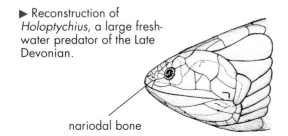

nariodal bone

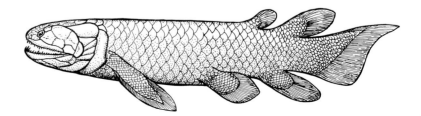

teeth in parallel rows. The whole dentition of porolepiforms reflects their predatory diet—large teeth or fangs appear regularly along the lower and upper jaws, flanked by several series of smaller gripping teeth. The robust ventral gill arch bones also had many small bones bearing tooth-like denticles.

One of the more widespread Middle Devonian holoptychioids was *Glyptolepis*, known from several species found in East Greenland and Scotland, and the subject of a very detailed study by Professor Erik Jarvik of Stockholm. *Glyptolepis* thrived in the Middle and early Late Devonian of the Old Red Continent and may have reached sizes close to 1 m in length. Recently some material of so-called *Glyptolepis* and *Holoptychius* species from Scotland has been re-described by Dr Per Ahlberg of the Natural History Museum, London as new forms, such as *Duffichthys*, from the Scat Craig Beds of Elgin. *Duffichthys* is known only from lower jaws, which are unusual in their very large attachment area for the symphysial tooth whorl.

The largest and one of the more widespread members of the porolepiform group was *Holoptychius*, which lived near the end of the Devonian. The scales of the largest species, *Holoptychius nobilissimus*, indicate that the fish may have been up to 2 m long, making it a formidable predator in the ancient river and lake systems of North America,

Greenland, Europe, parts of Asia and Australia. Whole body fossils of *Holoptychius* are well known from the famous Dura Den site in Scotland, where schools of these fishes died and were rapidly buried by wind-blown sands. On average the Dura Den *Holoptychius* are rather small species less than 1 m in length. *Holoptychius*-type scales have been found around the world, indicating the genus was widely dispersed and capable of transgressing saltwater to invade new river systems. Marine porolepiforms are known from Latvia (for example *Laccognathus* from the Lode deposit—see page 182), although in general they are rarely found outside of river or lake deposits.

THE DAGGER-TOOTHED FISHES (ONYCHODONTIFORMES)

The Onychodontiformes are a poorly known group of Devonian crossopterygians, which feature lower jaws with large dagger-like tooth whorls. Until the discovery of the well-preserved *Onychodus* specimens from the Gogo Formation of Western Australia, the group was known only from whole specimens of a little fish called *Strunius*, from Germany, and an assortment of partial skulls, jaws and bones belonging to other species in the genus *Onychodus*, which is now known to have been widespread around the Devonian world. The detailed study of the Gogo specimens by Dr S.M. Andrews of the National Museum of Scotland is still under way, so only preliminary results are able to be discussed here.

The largest species of *Onychodus* grew to about 2 m or so in length and may have been a lurking predator, much like today's moray eels in reef habitats. The skull was very kinetic, with a large hinge between the two divisions of the braincase, that enabled easy movement when the snout was raised. This is clearly an adaptation to allow the large fangs of the lower jaw to the fully utilised. Indeed, when the mouth of *Onychodus* was closed, the large lower jaw fangs almost touch the skull roof.

Other features of onychodontids' skulls that make them different from other crossopterygians are in the pattern of skull-roof and cheek bones. The upper jaw bone, or maxillary, is much like that of an actinopterygian in its general shape in having a large postorbital blade. The cheek features two large almost equally-sized bones, the squamosal and preoperculum;

◀ Reconstruction of *Strunius*, a little onychodontid from the Middle–Late Devonian of Germany.

▼ Reconstructed scenario based on an actual specimen of *Onychodus* from Gogo, which had swallowed a placoderm half its length.

John A. Long

and three infraorbitals flank the eye. The opercular mechanism is typical for crossopterygians, except that the submarginal bones may be absent. The body of *Onychodus* is long and slender, and from limited fossil evidence it appears that the dorsal fins were placed far at the back near the tail. The pectoral and pelvic fins are not well known from fossil material, although ossification of the humerus indicates that at least the front paired fins were quite robust with powerful muscular attachments.

One specimen of *Onychodus* from Gogo actually shows the remains of its prey. The bones of a small placoderm were found lodged in the position of the throat of the partial *Onychodus* skull. As the placoderm bones were facing forwards (away from the direction of the *Onychodus*) and these were entirely intact, showing no signs of damage, it has been deduced that the prey was captured by the tail and then swallowed whole. The measurements show that the placoderm would have been about 30 cm long whereas the *Onychodus* was about 60–70 cm in length, thus capable of catching and gulping live fishes half its own size!

GIANT KILLERS OF THE CARBONIFEROUS: RHIZODONTIFORMS

The rhizodontiforms (meaning "root tooth", so-named because of long fangs that extend deeply into the jaws) were the largest and most voracious of the crossopterygian fishes, reaching estimated sizes of up to 6–7 m. As they are generally known only from large pieces, we had little information about the group until the first complete material was described in 1986, and the first detailed study of a complete skull published in 1989. In general the group is characterised by stiff fins, which have long, unbranched bony rods (lepidotrichia) supporting the main part of the fin. The pectoral fins were very strong, supported internally by a robust humerus and strong ulna and radius bones. This pattern of arm bones is also seen in osteolepiforms and all higher land vertebrates.

The shoulder joint in rhizodonts may have been capable of powerful rotational movements, so that a fish could use its large, stiff fins to twist around in the water, similar to the way in which crocodiles tear flesh off their prey. The teeth of some large rhizodontiforms are laterally compressed to form a razor-sharp blade edge, also a characteristic feature of the group.

Dr Gavin Young, Canberra

▲ *Notorhizodon*, the earliest rhizodontid, may have reached lengths in excess of 3 m. This specimen shows part of the lower jaw with an abnormally folded over tooth.

▶ The skull of *Barameda decipiens*, from the Lower Carboniferous of Mansfield, is one of the only complete heads so far described for the Rhizodontiformes.

John A. Long

The rhizodontiforms first appear in the Middle Devonian, represented by *Notorhizodon*, discovered near the top of Mount Ritchie in South Victoria Land, Antarctica, in the early 1970s. *Notorhizodon* (meaning "southern root tooth") is known from a partial skull and lower jaws, and parts of the pectoral girdle. It was the largest predatory fish in the ancient Devonian Antarctic fish fauna, reaching an estimated size over 3 m. The only other known Devonian rhizodont is *Sauripteris* from the Late Devonian of North America, known principally from a large fossil pectoral fin skeleton and some other bits and pieces. The size of this fin indicates a large fish of similar size to *Notorhizodon*.

During the Carboniferous the rhizodontiforms reached a peak of diversity and size. The largest known form was *Rhizodus* from Scotland. An isolated lower jaw in the National Museum of Scotland is almost 1 m long, suggesting a maximum size of 6–7 m for the owner of the jaw. The largest teeth are the fangs at the front of the mouth, some of which are 22 cm long. Other forms, such as *Strepsodus*, *Barameda* and *Screbinodus*, also reached large sizes, and had similar large fangs at the

Dr Marius Arsenault, Parc de Miguasha

John A. Long

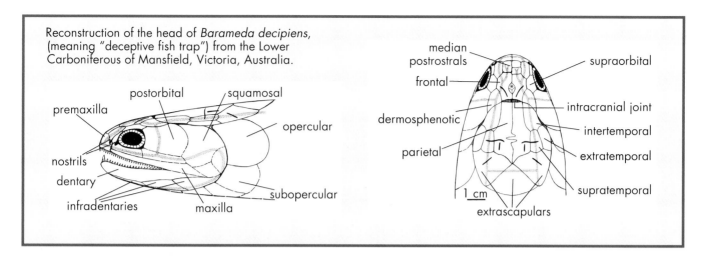

Reconstruction of the head of *Barameda decipiens*, (meaning "deceptive fish trap") from the Lower Carboniferous of Mansfield, Victoria, Australia.

premaxilla · postorbital · squamosal · opercular · nostrils · dentary · infradentaries · maxilla · subopercular

median postrostrals · frontal · dermosphenotic · parietal · supraorbital · intracranial joint · intertemporal · extratemporal · supratemporal · extrascapulars · 1 cm

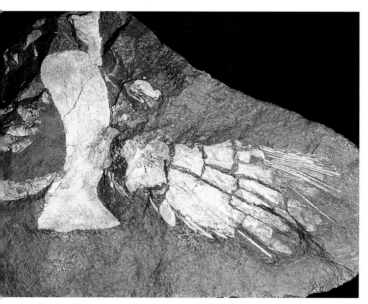

▲ A pectoral fin and shoulder girdle of the rhizodont *Sauripteris halli* from the Late Devonian near Blossburg, Pennsylvania, USA. Note the humerus articulating with the ulna and radius, the same pattern of bones seen in all tetrapods.

front of the lower jaws. *Barameda*, from Mansfield, southeastern Australia, shows that the head of rhizodontiforms is of similar pattern to that in some osteolepiforms and that the cranial joint enabled great frontal lift of the snout when opening the mouth. The rhizodontiforms were probably hunters of the large amphibians and fishes that lived in the murky coal swamps and lakes of their time.

The only complete fossil of a rhizodontiform is a small specimen of *Strepsodus*, whose body was relatively elongated with small pelvic, dorsal and anal fins, and large paddle-like pectoral fins. This body form is ideal for a slow-swimming stalker, capable of occasional fast bursts of activity as unsuspecting prey swim near by. The powerful jaws would then hold firm the struggling victim while the powerful paddle-like fins would enable the fish to twist and writhe in the water, tearing the prey apart and swallowing large chunks of meat. The last rhizodontiforms died out by the start of the Permian Period, probably outcompeted by the rapidly growing number of large aquatic amphibians sharing their habitat.

A STEP TOWARDS LAND: THE OSTEOLEPIFORMES

The Osteolepiformes were a diverse group of crossopterygians that first appeared in the Middle Devonian and became extinct during the Permian. They and the panderichthyids are the only groups of crossopterygians that possess a single external pair of

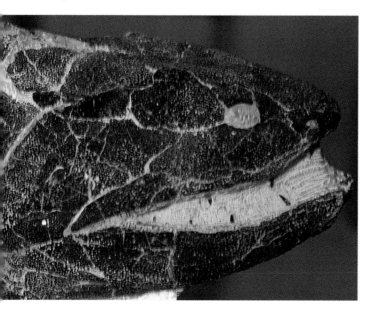

◄ Fossil crossoptergian fish, *Eusthenopteron*, one of the Osteolepiformes, from the Late Devonian Escuminac Formation, Quebec. *Eusthenopteron* is perhaps one of the most intensely studied of all fossil fishes. Its anatomy was reconstructed by the Swedish palaeontologist Professor Erik Jarvik by making painstaking wax models built from a specimen ground away in layers of one-tenth of a millimetre thick. The work took nearly 30 years.

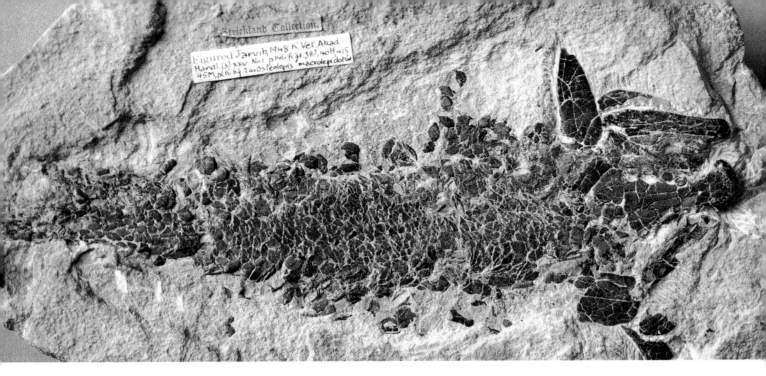

▲ *Osteolepis*, one of the primitive cosmine-covered osteolepiform fishes, of Middle Devonian age, from the Old Red Sandstone of Scotland.

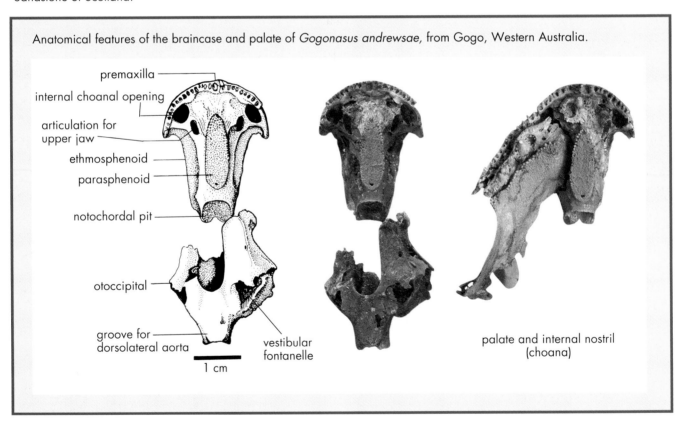

Anatomical features of the braincase and palate of *Gogonasus andrewsae*, from Gogo, Western Australia.

premaxilla
internal choanal opening
articulation for upper jaw
ethmosphenoid
parasphenoid
notochordal pit
otoccipital
groove for dorsolateral aorta
vestibular fontanelle
1 cm

palate and internal nostril (choana)

nasal openings and a choana or palatal nostril, and generally have a set pattern of seven bones forming the cheek unit. Like rhizodontiforms they have strongly ossified paired fins, the pectoral fin having a solid humerus, ulna and radius. The most primitive members of the group have thick rhombic scales and all dermal bones have a cosmine layer, although in several later lineages the cosmine is lost and the scales become thinner and rounded.

The earliest group of osteolepiform fishes are the family Osteolepididae, represented by well-known forms such as *Osteolepis*, *Gyroptychius* and *Thursius* from the Old Red Sandstone beds of Scotland. These are generally small fishes, less than 50 cm long, and relatively conservative in their patterns of dermal bones. All have simple heterocercal tails and two dorsal fins.

Megalichthyinids were a group of advanced

osteolepidids that survived until the Middle Permian, long after all other families had died out. They retained their cosmine cover, but were specialised in having special tectal bones wrapping around the nostril, wide palatal cavities and a long median process extending into the mouth from the premaxillary bones. Some megal-ichthyinids also possessed a special articulation on the dermal bones linking the two halves of the skull roof. The group were highly successful and inhabited the coal swamps and lakes around Gondwana and also in Euramerica, reaching sizes of about 1 m or so.

The Eusthenopteridae include forms with reduced cosmine cover, and most have lost the extratemporal bone from the skull roof, except for one primitive genus, *Marsdenichthys*, from Australia. The best-known member of the family is *Eusthenopteron foordi* from the Late Devonian Escuminac Bay fauna of Canada, a medium-sized fish about 1 m long. *Eusthenopteron* was the subject of 25 years' detailed study by Professor Erik Jarvik, of Stockholm, who made a large wax model of its braincase and gill arches from serial sections of a skull. The skull was embedded in resin, then ground down by a tenth of a millimetre at a time. The section was then photographed and drawn, magnified by ten, and a wax layer was cut out to show the position of bone and cartilage. By slowly building up the wax layers Jarvik constructed his model and then published several large papers elucidating the fine anatomical structure of *Eusthenopteron*. Today this genus is still one of the best known of all Palaeozoic fishes. Complete fossils of *Eusthenopteron* are frequently found in the Escuminac Formation of Quebec, and other species of the genus have been described from Russia and Europe.

John A. Long

▲ The lower jaw of *Rhizodus hibberti*, from the Early Carboniferous of Scotland. The largest jaw measures nearly 1 m long with massive fangs in the front of the jaw up to 22 cm high, suggesting the maximum size of 6 to 7 m.

There are several other groups of Osteolepiformes that can be mentioned, although most differ in only technical ways from the general plan of the group. Among the recently discovered new groups of Osteolepiformes is the family Canowindridae, a group unique to the East Gondwanan region (Australia and Antarctica). Characteristics of this primitive group of Osteolepiformes include large deep opercular bones, wide flat skulls with very small eyes and there may be additional bones occupying the position of the postorbital bone in the cheek units. The first described genus, *Canowindra grossi* (named after the New South Wales town of Canowindra) was studied by Dr Keith Thompson in 1973 but could not be attributed to any particular group of crossopterygians at that time. In recent years additional genera have been described from Victoria (*Beelarongia*) and Antarctica (*Koharolepis*), filling in the evolutionary sequence of the group. The most primitive members have rhombic cosmine scales and bones, whereas the last genus, *Canowindra*, lacks cosmine and has rounded scales with a ventral boss on them, as in eusthenopterids.

Osteolepiformes are the most important group of fishes in the evolutionary transition from fish to land animal, particularly the panderichthyid group. These fishes have more in common with primitive amphibians than any other fish. The great step in vertebrate evolution is discussed in more detail in the last chapter of this book.

▼ Reconstructed scene of an Early Carboniferous lake with a gigantic *Rhizodus* catching an unwary amphibian.

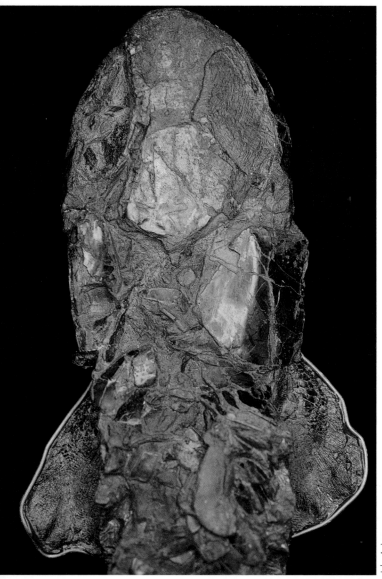

John A. Long

John A. Long

◄ Skull and front half of body of an Early Carboniferous megal-ichthyinid fish from central Queensland, viewed from above.

Dr Oleg Lebedev, Palaeontological Institute, Moscow

▲▲ (top) The skull roof of *Megalichthys laticeps*, from the Carboniferous of Scotland. Note the shiny cosmine surface.

▲ Reconstruction of the head region of *Koharolepis*, a Middle Devonian canowindrid osteolepiform from Mount Crean, southern Victoria Land, Antarctica.

◄ The skull of *Megapomus*, a Carboniferous megalichthyinid fish from Russia.

THE ORIGIN OF PAIRED LIMBS

The oldest fish fossils are completely devoid of paired fins, although some early jawless fishes soon acquired simple fin folds along the sides of the body. The precursors to the arms and legs of higher vertebrates are the pectoral and pelvic fins and their support structures (internal shoulder and hip girdles). This vital step in vertebrate evolution is actually one achieved partly within the jawless fishes and completed by the gnathostomes.

The most primitive fishes to possess internal shoulder girdles that supported a pectoral fin are the armoured osteostracans. These have a primitive scapulocoracoid bone with two or three perforations for vascular and nerve supply to the fin. In many forms the head shield is well developed around the shoulder girdle, forming a "cut-water" for the pectoral fin. The pectoral fins in these agnathans appear to lack ossified fin bones, and most likely had a few simple cartilaginous fin rays. Thus the earliest arms or pectoral fins appear to have lacked any significant degree of muscular control, and probably functioned by simple up-and-down movements, which aided the steering of the front of the fish. None of the agnathans show any development of pelvic fins or girdles; these were to come close on the heels, so to speak, of the evolution of jaws.

The oldest gnathostome fishes that are well preserved in entirety are the placoderms and acanthodians, both of which show the presence of well-developed pectoral and pelvic girdles. Placoderms preserved in three-dimensional form confirm that in arthrodires the pelvic girdle is a perichondrally ossified bone with numerous foramina for the passage of nerves, arteries and veins. In some species, simple perichondral tubes formed around the pelvic fin rays.

A pectoral fin and shoulder girdle of *Barameda decipiens*, from Mansfield, Victoria, Australia. These robust bones are the commonest fossil remains of rhizodontiform fishes.

caput humeri
humerus
entepicondylar process
1 cm
axial mesomere bone
ulna
radius
unjointed fin rays (lepidotrichs)

Placoderms also show the earliest development of sexual dimorphism: pelvic girdles with male intromittent organs (claspers), as seen in one group, the ptyctodontids (see page 115 for discussion).

The evolution of the pectoral fin to become the standard vertebrate arm pattern can be seen in a series of stages from chondrichthyans to simple ray-finned fishes to lobe-finned crossopterygians. Primitively the pectoral fin has many rays supporting it, and these are grouped into three main regions: a leading propterygium, a middle division called the mesopterygium, and a branching third region, the metapterygium. Even in some very primitive jawed fishes, such as sharks, the metapterygium can be seen to branch in such a way that a leading solid bone articulates with two stout elements, which then branch further down the fin. It is the loss of the first two parts of the fin, and the greater development of the metapterygium, that results in the robust crossopterygian lobe fin. The mechanism for this to have occurred was heterochrony: the paedomorphic loss of the pro- and mesapterygium, accompanied by the peramorphic development of the metapterygium, eventually displacing the first two divisions of the fin entirely.

In the advanced crossopterygians such as osteolepiforms, the first metapterygial element is called the humerus and is the same bone as that supporting the upper arm in all later tetrapods. It is a robust, complex bone with a caput humeri and well-developed entepicondylar process and foramen, features otherwise seen only in tetrapods. The humerus in these fishes articulates with an ulna and radius, but after that there is no one-to-one correspondence with the "wrist" bones of primitive tetrapods.

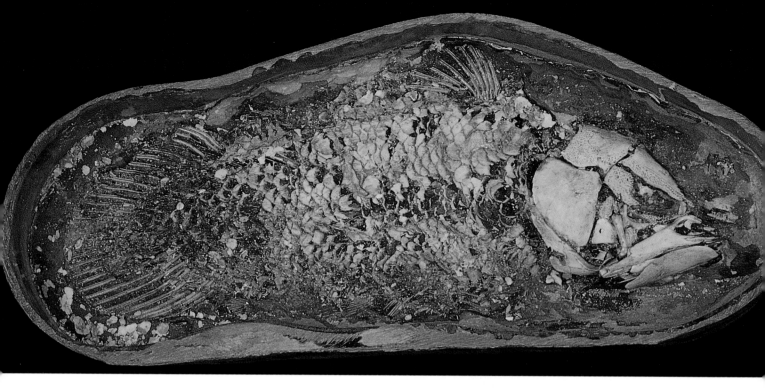

An Early Cretaceous coelacanth, *Axelrodichthys*, from the Santana Formation of Brazil.

▼ *Callistipterus*, a Late Devonian eusthenopterid fish from the Escuminac Formation of Quebec, Canada.

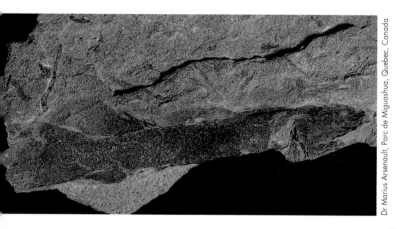

Dr Marius Arsenault, Parc de Miguashua, Quebec, Canada

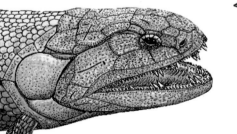

▶ Reconstruction of *Canowindra grossi* from the Late Devonian Mandagery Sandstone of central New South Wales, Australia.

▲ Reconstruction of *Marsdenichthys* from the Late Devonian of Victoria, Australia. *Marsdenichthys* was one of the largest predators in the Mount Howitt lake ecosystem.

RELATIONSHIPS OF THE CROSSOPTERYGIAN GROUPS

Relationships of the crossopterygian groups are hotly debated by vertebrate palaeontologists, and consequently a number of differing opinions exist in the current literature. One school of thought, led by British palaeontologists Peter Forey, Brian Gardiner and Colin Patterson, argues that lungfishes could be the closest ancestors to the land vertebrates and that lungfishes are at the pinnacle of osteichthyan evolution. This implies that the lungfishes are a subgroup within the crossopterygians, and it has been criticised in recent years by a number of workers on early osteichthyan fossils. One flaw in the scheme is that many of the characteristics chosen to show that lungfishes are more advanced than crossopterygians are based on the soft tissues of living fishes and amphibians and cannot be tested against the many extinct groups of crossopterygians, but only against the coelacanth *Latimeria*.

Furthermore, recent finds of extremely well-preserved crossopterygian fossils from Russia and Australia have been shedding new light on the anatomical features of the group, and evidence is

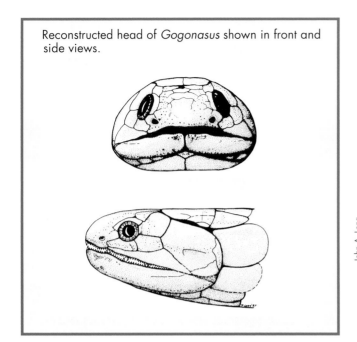

Reconstructed head of *Gogonasus* shown in front and side views.

▲ Close up of the head of *Canowindra grossi* from the Late Devonian of New South Wales, Australia. Note the very small eyes and clear intracranial joint. This specimen is a latex cast whitened with ammonium chloride. The scale is 1 cm.

mounting in favour of the traditional viewpoint, held by nineteenth-century palaeontologists, that the crossopterygians most likely gave rise to the first amphibians. In this book I regard lungfishes as a group distinctive from crossopterygians and not closely related to tetrapods.

One scheme of crossopterygian interrelationships first proposed by Professor Hans-Peter Schultze, of Kansas, in 1987, and since modified slightly by me, recognises the crossopterygians as a natural grouping of fishes, with successive stages in their evolution clearly delineated. The scheme recognises two major divisions within the crossopterygians, one containing the "primitive" orders of crossopterygians (such as the coelacanths and onychodontids) and the other being the Rhipidistia (Porolepiformes, Rhizodontiformes, Osteolepiformes, Panderichthyida and Tetrapoda). Rhipidistians have well-developed skull roofs with a set pattern of bones, lower jaw with four large infradentaries, plicidentine (infolded enamel and dentine in a complex pattern) in the teeth, and a sensory-line canal on the skull in which the supraorbital canal joins the main lateral line. Recent research by Drs Per Ahlberg and Richard Cloutier has shown that the so-called "stem-group

◄ *Beelarongia*, a Late Devonian canowindrid osteolepiform from Mount Howitt, Victoria. These fishes are thought to be the most primitive of all known osteolepiformes.

▼ *Marsdenichthys longioccipitus*, a primitive Late Devonian eusthenopterid from Mount Howitt, Victoria, Australia. This is a whitened latex peel of the skull viewed from above.

crossopterygians" *Powichthys* and *Youngolepis* show intermediate stages in the closure of the intra-cranial joint, a condition indicating that they are actually more closely related to lungfishes. This theory would place lungfishes at the pinnacle of one lineage of crossopterygian evolution.

The most specialised subgroup of rhipidistians includes those forms having a strong pectoral fin with humerus, ulna and radius present (Rhizodonti-formes, Osteolepiformes, Panderichthyida and Tetrapoda); and the penultimate group contains those forms with a palatal nostril, or choana (Osteolepiformes, Panderichthyida and Tetrapoda). The close affinities between the panderichthyid fishes and the first amphibians has only just been elucidated in a series of recent scientific papers and will be discussed in depth in the final chapter.

Chapter 10

THE GREAT STEP IN EVOLUTION

FROM FISHES TO LAND ANIMALS

The greatest step in vertebrate evolution is undoubtedly the transition from aqueous gill-respiring fishes to air-breathing, walking, land animals. Despite the numerous complex anatomical changes that this requires, the transition from osteolepiform fishes to primitive amphibians was no great deal, as already the prerequisites for the change were inherent in the group. The limbs, shoulder and hip girdles of these fishes were strong. Their arms and legs (as fins) already had the same bone pattern that all land animals would retain. Lungs were already present as swim-bladders, and the pattern of skull and cheek bones did not have to change at all. Not long after the first of the panderichthyid fishes appeared in the early Late Devonian, their lineage had gone one step further—the ultimate step that was to guide the future trend of all subsequent vertebrate evolution. From panderichthyid-like fish arose primitive amphibians. These early land-lubbers did not conform to the standard tetrapod pattern and retained many fish-like characters, such as gill breathing, many fingers and toes on the limbs, a fish-like tail, and scales covering much of the body. Within a relatively short time after the first amphibians began to diversify, the first proto-reptile had appeared, in the Early Carboniferous. The rest of vertebrate evolution is simple by comparison, being merely a tale of what lineages stemmed from the basic reptilian pattern.

The transition from fishes to four-legged land vertebrates (tetrapods) has always been a difficult area of evolution for the lay person to grasp and has, until quite recently, been a hotly disputed area of palaeontology, particularly with regard to which fishes may have given rise to the first amphibians. The following passage taken from an early twentieth-century popular geology book nicely invokes some of the mystery surrounding this major step in evolution:

These facts, coupled with the fish-like structure of certain genera [of fossil amphibian], tempt one to imagine them as having slowly evolved, in the midst of the Carboniferous marshes, from true fishes; first wriggling helplessly among the slime, and afterwards generally acquiring lungs for breathing air and limbs for locomotive purposes, and lastly, their strong and peculiar teeth for masticating the vegetation on which they may be presumed to have principally lived. (from B. Webster Smith, The World in The Past, *Warne & Co., London, 1926)*

▶ The skull of *Acanthostega*, one of the earliest complete fossil amphibians, of Late Devonian age, from East Greenland.

Previous page: The panderichthyids were fishes that closely approached tetrapods in their overall anatomy. This is the skull, viewed from above, of *Panderichthys rhombolepis* from the Late Devonian Lode site, Latvia. *Dr Oleg Lebedev, Palaeontological Institute, Moscow*

Inset: Professor Erik Jarvik in east Greenland, showing the outcrops of Late Devonian sandstones that have yielded the earliest well-preserved amphibian remains. *Hans Bjerring, Swedish Museum of Natural History*

Dr Mike Coates, Cambridge University

THE INVASION OF LAND

Animals and plants invaded land for the first time in the Silurian Period, well before fishes ever ventured a fin upon the sandy riverbanks. The earliest indirect evidence of life on land is from plant spores occurring in the late Ordovician Period, about 470 million years ago. These resemble the spores of very primitive plants alive today, such as liverworts and mosses. However, the first good record of vascular plants comes from the Late Silurian, about 420 million years ago. These consist of plants with thick stems and spiny leaf-like structures, looking very sedge-like in most respects. One of these, *Baragwanathia* from Australia, resembles a club moss, and its remains often occur in marine deposits, indicating it grew close to the seashore.

About 400 million years ago we have the first good association of land animals and plants forming a simple terrestrial ecosystem, preserved in the Rhynie Chert of Aberdeenshire, Scotland. The animals include spider-like trigonotarbids, mites, springtails, and shrimp-like creatures, which lived alongside simple vascular plants such as *Rhynia* and *Algaophyton*. The environment of Rhynie has been interpreted as a series of hot springs in a volcanic setting, and the image conjured up is one of the land being only slightly green, still mostly sparse, with poorly developed soils but hosting a community of very small invertebrates (all only a few millimetres long).

An assemblage of trace fossils and rare body fossils from the Tumblagooda Sandstone, near Kalbarri, Western Australia, indicates that in the Late Silurian much larger invertebrate ani-mals were venturing onto land. Large sea-scorpions (eurypterids), some nearly a metre long, and primitive multi-segmented arthropods such as euthycarcinoids, the possible ancestors of the first insects, inhabited these riverplain environments, along with a great diversity of creatures known only from their footprints, feeding traces and burrows. These indicate that at least 15 different creatures lived in this environment.

By the start of the Devonian we see a dramatic increase in the diversity of both land plants and

John A. Long

▲ Trackways made by sea-scorpions (eurypterids) that walked on land about 420 million years ago near Kalbarri, Western Australia. Note how the clear tracks, right, made on firm sands exposed to the air, fade as they go left, where a pool of water stood, and where the marker pen now rests. These constitute the oldest recorded evidence of life invading land yet found.

▼ A Devonian fern. Although plants began invading the land in the late Silurian, the first large forests developed during the Devonian, at much the same time as vertebrates first invaded the land.

Dr Marius Arsenault, Parc de Miguashua

invertebrate life. Large plants, like the horse-tails (lycopods) reached enormous sizes, tens of metres high, and formed the first real forests, along with simple ferns, psilophytes, and lower plants such as worts, mosses and sedges. Fungi also thrived in the primitive terrestrial environments. Large land scorpions, spiders, many kinds of mites, centipede-like animals, millipedes and worms were all living on the land by the Middle Devonian—the time when the first fishes were acquiring the ability to breathe air and may have been lured into the now-established land-based ecosystem. All these land-based Devonian invertebrate communities were carnivorous, feeding on each other. The oldest evidence of herbivory is from the Carboniferous Period, where fossil leaves show undoubted damage caused by insects (it has been suggested that, before this, insects had not evolved the ability to digest lignin in plants).

Thus the land was well colonised by plants and invertebrates by the Middle Devonian, setting the stage for the greatest step in vertebrate evolution as fishes began venturing onto land and discovering a new world opening up to them.

Thus early twentieth-century scientists saw land animals as springing from fishes wriggling in the mud, which, being out of water long enough, somehow acquired lungs and limbs more or less at the same time. Yet we now know that fishes already had lungs well before they ventured from the water, and that certain fishes had the same set of arm and leg bones as land animals. In fact most of the major transitions needed for the invasion of land had already occurred within fish evolution. Today we have a far more complete record of early fossil amphibians and more detailed knowledge of the physiology of living air-breathing fishes, such as the dipnoans. More significantly, we can study the nature of growth and change in organisms and document the major morphological transformations that result from simple changes in the timing of a creature's embryonic or growth development (termed heterochrony). By combining all of these new results and discoveries, an accurate picture of the origins of land vertebrates is beginning to emerge.

The discovery of fossil amphibians that lived in the Carboniferous Period goes back a long way in palae-ontology. However, it was only in the twentieth-century that the first remains of amphibians from Devonian rocks were discovered. Furthermore, only in the past decade have there been many new finds of

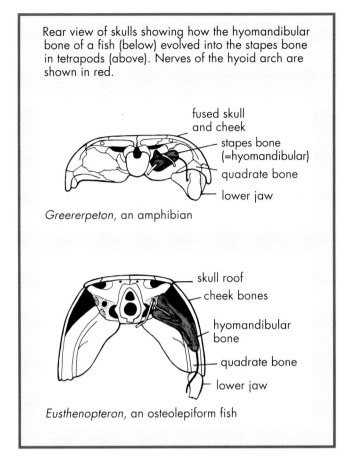

Rear view of skulls showing how the hyomandibular bone of a fish (below) evolved into the stapes bone in tetrapods (above). Nerves of the hyoid arch are shown in red.

fused skull and cheek
stapes bone (=hyomandibular)
quadrate bone
lower jaw

Greererpeton, an amphibian

skull roof
cheek bones
hyomandibular bone
quadrate bone
lower jaw

Eusthenopteron, an osteolepiform fish

◀ An Early Carboniferous tetrapod skull from the East Kirkton Limestone, near Bathgate, Edinburgh. These same beds have yielded the oldest remains of reptiles.

▼ Close up of the snout of *Panderichthys rhombolepis* from the Lode site, Latvia.

well-preserved Devonian amphibians and their trackways that have shed much new light on the remarkable transition between fishes and the tetrapods. Such discoveries show that the first tetrapods were not very different from their fishy ancestors. They may have had limbs with digits, as opposed to fins, but some remained fully aquatic animals, capable of underwater gill respiration, and probably spent most of their life living like fishes.

Dr Oleg Lebedev, Palaeontological Institute, Moscow

FROM WATER TO LAND: HOW TO SURVIVE

Life in the water is quite different from life on land. Fishes have a different body shape from land animals because the water holds them up, and the forces acting on their bodies are different. Thus gravity is one force, friction from the water is another, and the buoyancy of their bodies within the water column is yet another force. By balancing its buoyancy, a fish can achieve neutral weight in the water (that is, weigh nothing) and so can direct its energy from the tail and hydrodynamic lift from fins to take it up or down in its environment. In reality, however, it is not as simple. Many fishes (osteichthyans) use a swim-bladder to regulate gases inside the body to achieve lift or fall in the water through subtle changes in buoyancy. Sharks, and other fishes without swim-bladders use different methods to achieve neutral buoyancy, like having a large oil-filled liver and having wing-like pectoral fins to give greater lift in the water.

But there is more to life in water than just supporting your body and going up and down. There's also breathing, sensing, eating, excreting, and trying to avoid being eaten, not to mention the most important of all functions, reproduction. All of these functions needed to be modified for a successful conquering of the terrestrial environment.

Breathing through gills requires that there is always a certain sustainable level of oxygen dissolved in the water so that the oxygen can be taken up through the fine membranes of the gills into the blood stream. In order to keep a constant flow of water over the gills, fishes use different pumping methods or, as in sharks, tend to keep moving or rest in active currents of water. Thus water comes in through the mouth and either flows over the gills and out via the operculum or is pumped over the gills by moving the mouth and altering the volumes inside the gill chamber. The transition to air breathing was not difficult in this regard. Many modern gill-breathing fishes also have an ability to breathe air for limited times, using the swim-bladder as a gas-exchange organ to obtain small

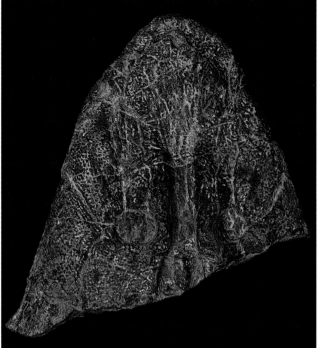

Dr Marius Arsenault, Parc de Miguashua

▲ The skull roof of *Elpistostege*, a panderichthyid fish from the Late Devonian Escuminac Formation, Quebec, Canada. When first discovered the skull roof of this fish was described as belonging to an early amphibian.

◄ Reconstruction of *Panderichthys*. Note the paired fins (limbs) and absence of other fins, and the relatively large head with eyes on top of the skull.

amounts of oxygen (for example, mudskippers).

Eating may seem to be a basic requirement of all creatures and may not necessarily have been a driving force for fishes to leave the water. Indeed the first land vertebrates were of rather clumsy design and probably ventured away from the water for only short periods. Their crossopterygian-like bodies and skulls were much better suited to catching prey in the water. Eating food on land, and acquiring the body design to do this efficiently, came much later in the evolution of the tetrapods. In order for a slow-moving amphibian to catch land-living insects as a source of food, one of two requirements is needed. Either the animal must be capable of short fast lunges or bouts of running to catch the unsuspecting prey; or, as in many modern frogs, they must develop a long prehensile tongue that can simply dart out and catch a flying insect. We have no evidence that the long tongue existed in early tetrapods and is more likely a specialisation of later amphibians, so must therefore rely on evidence from their body shapes and limb joints that they were not efficient hunters on land.

Sensing the environment around you is another important part of invading a new habitat. In the water, fishes rely heavily on their lateral-line sensory system for detecting movements in the water, either to find prey or to detect a larger predator approaching them. They also use their eyes and hearing to lesser extent. Fishes that rely on leaving water for short periods of time rely even more on their eyes for detecting food items or approaching danger. One of the major differences between the bones of fishes and those of fossil tetrapods is that fishes tend to have the lateral-line canals enclosed with rows of pores open to the external environment, whereas the tetrapods have wide, open grooves in the dermal bones for their sensory lines. Eventually as these animals became more adapted to life on land, the other senses took over from the lateral-line system, which is really only useful when the creature is in water. One of the first major transitions to occur in this respect was the evolution of the eardrum in early amphibians. The long hyomandibular bone that braces the jaw joint of crossopterygian fishes is, in tetrapods, modified to brace the otic membrane, or skin above the inner ear. Thus the stapes bone evolved as a means to convey airborne vibrations into the inner ear. The evolution of more complex middle ear bones, such as the incus and malleus, came later with the reptiles and mammals. Although the eyes and nostrils became increasingly more important for the land animal to sense its surrounding habitat, there is virtually no initial change in these structures from the fish to the first tetrapod.

Excreting is another vital body function and is linked with the overall problem of preventing the animal from dehydrating once out of its aqueous environment. Fishes and water-dwelling animals that excrete into the water do not normally risk dehydration, as they are always immersed in water. However, a land animal that excretes urine and moist faeces is continually lowering its level of body moisture and must replenish that supply by drinking more. Early amphibians had skins that were covered with fish-like scales to protect the skin and prevent some dehydration. The first real adaptation for living away from the water was to come with the reptiles, only a short time after the first amphibians appeared. This great innovation was the hard-shelled egg, and this testifies that the the egg-layers were the first creatures capable of living away from the water for any significant length of time. Yes, this does indicate the true answer to the age old riddle — the egg came well before the chicken.

PANDERICHTHYIDS: FISH, OR AMPHIBIANS WITH FINS?

The last few chapters in this book have given a good overview of the lobe-finned fishes, both the lungfishes and the crossopterygians. However, one very special group, the panderichthyid fishes, have been saved until last. Until quite recently they were always classified as a group within the osteolepiform crossopterygians, although a paper published in 1991 by German palaeontologist Hans-Peter Schultze working with Russian palaeontologist Emilia Vorobyeva has redescribed the group and placed these fishes in a new order, the Panderichthyida. The reason they have done this is because panderichthyids share more features in common with early amphibians than with any other fish group. So what exactly makes a panderichthyid?

The Order Panderichthyida contains only five species, belonging to three genera. Of these, only one, *Panderichthys* (meaning "Pander's fish"), is known in any detail. *Panderichthys* was first described from isolated snout and skull material by German scientist Walter Gross in 1941 from remains found in the famous Latvian site at Lode. Fossil material from this site is beautifully preserved in three-dimensional form, as the surrounding sediment was a soft clay. New finds of more-complete specimens of *Panderichthys* have enabled description of nearly all aspects of its anatomy by Emilia Vorobyeva and her colleagues.

Panderichthyids are characterised by the following few features: the median rostral bone of the snout does not contact the premaxilla; they have a very large median gular bone under the head; the mouth is underneath the protruding snout; and they have a lateral recess in the nasal capsules.

They share a number of specialised features with amphibians, as the following list demonstrates. They are long-bodied fishes with a large head that is nearly one-quarter their total length. They have large pectoral fins and smaller pelvic fins, but no dorsal or anal fins. The skull is broad and flat, with eyes on the top of the head and rather close together with distinct brow ridges. The external nostrils are placed ventrally, close to the margin of the mouth. A cross-section of the large teeth shows a complex form of labyrinthine infolding of the enamel and dentine as occurs in many amphibians. The bones of the skull roof have three pairs of median bones from the back of the skull to the eyes, rather than two as in other crossopterygians. The cheek bones are large, and the jugal bone separates the squamosal from the maxilla, as occurs in amphibians.

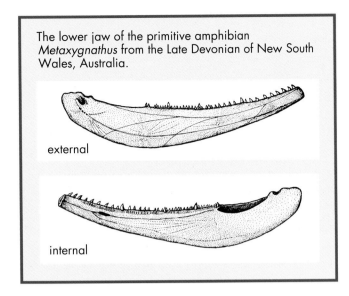

The lower jaw of the primitive amphibian *Metaxygnathus* from the Late Devonian of New South Wales, Australia.

external

internal

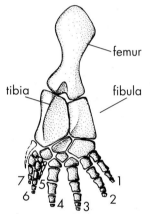

tibia

femur

fibula

7 5
6 4 3 2 1

◄ Restoration showing the hind limb of *Ichthyostega*, a Late Devonian amphibian with seven digits. (After the work of Dr Mike Coates and Dr Jenny Clack)

▼ The skull, viewed from above, of *Ichthyostega*, a Late Devonian amphibian from East Greenland.

John A. Long

The cheek bones do not rigidly meet the skull roof for some distance, leaving a large spiracular slit along each side of the skull table. The external pattern of skull roof bones shows that the intracranial joint has fused, so that the skull was not kinetic as in many other crossopterygians. The pectoral fins have strongly ossified humerus, ulna and radius, with the humerus having a longer shaft than for any other crossopterygian fish. The vertebrae are also unusual in having only the ventral component (intercentrum) present with large

neural arches straddling the notochord. Ribs are attached to the neural arch and intercentrum, exactly as in tetrapods. The body is covered in rhombic bony scales, but cosmine is absent.

Apart from *Panderichthys*, the second-best-known genus is *Elpistostege*, from the Escuminac Formation of Quebec. When British palaeontologist Stanley Westoll first described an incomplete skull roof of this creature in 1938, he was certain it was a primitive amphibian. It was not until new material was recovered in the 1980s that it was shown to be a panderichthyid crossopterygian, and its redescription lead Hans-Peter Schultze to pursue the close link between panderichthyids and tetrapods. *Elpistostege* is known from only a few partial skulls and is, in general terms, very close to *Panderichthys*. *Obruchevichthys*, the third genus in the group, is another Russian form of questionable assignment to the order, based on only a lower jaw that shows many features similar to early tetrapods. Some specialists prefer to regard it as another early tetrapod.

Thus the panderichthyids, when preserved as partial skull material, are almost impossible to distinguish from early amphibians, and this has led to some confusion in the past. Before we look at the sequence of events leading to the fish–amphibian transition, it is first necessary to see exactly what the earliest true tetrapods were like.

THE MOST PRIMITIVE AMPHIBIANS

The first amphibians known in considerable detail are the Late Devonian ichthyostegalids from East Greenland: *Ichthyostega* and *Acanthostega*. Early descriptive work by the Swedish scientists Gunnar Save-Soderbergh and Erik Jarvik revealed much about the basic structure of these amphibians, although recent work by the British team at Cambridge University, Mike Coates and Jenny Clack, is discovering many new aspects of the anatomy of these forms based on recent discoveries from East Greenland in the 1980s.

The skull is well ossified, and there is no sign of an intracranial joint as seen in crossopterygian fishes. Otherwise the skull roof and cheek pattern is much like that of a panderichthyid fish, with the exception of *Ichthyostega* which has an unusual fused median bone in rear of the skull roof. The head is large relative to the overall body size, and the eyes are placed in the middle and on top of the skull. The single pair of external nostrils opens close to the mouth and faces downwards. Inside the mouth there is a large palatal nostril (choana) present, bordered by the vomer and other toothed palatal bones. The braincase is not hinged and is reduced in size relative to skull overall. Recently the gill arch elements of *Acanthostega* were described, revealing that the animal was capable of aquatic respiration in the adult phase. Thus these early amphibians were still highly dependent on living in the water.

The bodies of these early amphibians show long tails that bear a well-developed tail fin supported by rods of bone (lepidotrichs). Scales looking like thin slivers of dermal bone may cover the ventral or belly surface. The limbs feature many digits (seven or eight) on the front and hind limbs, and these may be divided on each hand or foot into a series of large digits and series of much smaller elements. Another Devonian amphibian, *Tulerpeton*, described from Russia, has six digits on its front limb. There are no external shoulder girdle bones seen on the outside of these animals, although the cleithrum is a large, high bone.

In general the structure of these earliest amphibians is much like that already described for panderichthyids. Only the tetrapods have limbs with digits present while the fish have fins and more complete series of operculogular bones which cover the gill chamber.

WHERE DID THE FIRST TETRAPODS EVOLVE?

Where have the earliest tetrapod remains or traces come from? The Late Devonian is divided into two stages, the oldest being the Frasnian, the youngest being the Famennian. Apart from the East Greenland forms, of late Famennian stage, other skeletal remains of Devonian tetrapods have been found in Russia and the Baltic (*Tulerpeton, Ventustega;* also late Famennian), and from Scotland (possible tetrapod remains, late Frasnian, from Scat Craig), and from Australia (*Metaxygnathus*; middle Famennian). Footprints of Devonian amphibians have also been described from Australia, from Famennian red

sandstones exposed on the Genoa River, eastern Victoria, and from similar rocks of possible early Middle Devonian age in the Grampian Ranges of western Victoria, Australia. Some doubt has been cast on these last trackways, although recent re-evaluation of the age of the Grampians sandstones, based on discoveries of thelodont scales, would place them as most likely being Middle Devonian age. Thus the earliest fossil evidence of tetrapods, anywhere in the world, is in Australia. The Scat Craig material from Scotland is of questionable tetrapod affinity and may alternatively belong with advanced panderichthyid fishes. Therefore the next earliest, and more reliable evidence, is probably the *Metaxygnathus* jaw, also from Australia.

This patchy distribution poses a problem when considering the close relationship between the panderichthyids and Devonian tetrapods. Panderichthyids at present have a Euramerican distribution with no evidence of them having existed in the ancient southern supercontinent of Gondwana. Thus it would be logical to assume that Euramerica was the most likely place for tetrapods to have evolved. However Gondwana appears to have been the centre for much of crossopterygian evolution. Its unique faunas containing stem-group osteolepiform fishes, canowindrids, earliest rhizodontiforms and primitive known eusthenopterids, would suggest that the evolution of higher crossopterygians fishes may have first taken place there. This data, considered with the fact that the oldest evidence of tetrapods also comes from Gondwana, leads me to favour Gondwana as the most likely place for this great step in evolution. The potential for new fossil fish and tetrapod sites in Antarctica and parts of Australia will one day hold the key to this great mystery. As new Devonian fish faunas are continually uncovered and described from Australia, perhaps the solution is close at hand, and may be solved with the next season's field programme.

Comparison between the heads in side view of the osteolepiform fish *Eusthenopteron* (right) and the primitive amphibian *Crassigyrinus* (left). Similar colours indicate homologous bones.

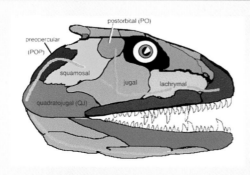

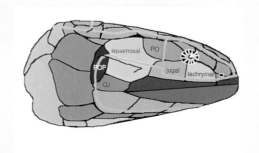

A SMALL STEP FOR FISHKIND,
BUT A GREAT STEP FOR MAN

We have now seen the contenders, the fishes, and the end result, the amphibians, and looked at where this great step may have occurred. But why? What factors drove fishes to eventually become land-lubbers? Was it simply a matter of escaping the piscine rat-race of the Devonian seas, rivers and lakes, or were there more fundamental reasons for invading a new, but hostile habitat? Well before this time land had been colonised not only by plants (with their great diversity), but also by many forms of arthropods and other invertebrates. No doubt there were new food sources to be exploited, but this was probably not reason enough to evolve digits and air-breathing capability. Like most innovations in evolution the precursor organs for the invasion of land were already installed in the fishes.

We have seen that the lungs of amphibians are just modified versions of the swim-bladder of an osteichthyan fish. The digits are just the ends of the fins, and already those fins in osteolepiforms and rhizodontiform fish had robust humerus, ulna and radius bones. The skulls of early amphibians and panderichthyid crossopterygians are almost identical, and little change was needed in this department. Recent work on the Late Devonian East Greenland amphibians shows that they were really just slightly modified fishes—they still inhabited the water and respired by gills with an assumed air-breathing capability in place, based on the presence of their internal palatal nostril or choana. The numerous digits on the hands and feet are closely akin to the paddle-shaped fins of their closest piscine ancestors. In other words all of the adaptations evolved in some early amphibians were probably developed as specialisations enabling them more efficiency in the water for their particular lifestyle. Many of the lobe-finned fishes may have been capable of venturing out of the water for short crawls or wriggles on the banks of the river. Maybe this was a way to escape and find quiet pools in which to lay eggs away from predators. Reproduction is often a powerful driving force in evolution, as obviously it is paramount to the continuation of the species.

One of the major considerations to the evolution of tetrapods, and one that has been largely overlooked in recent years, is the role played by changing rates of development, and how similarities in closely-related organisms may be disguised by different stages of maturity. A classic example of this is that the juveniles of the osteolepiform fish *Eusthenopteron* actually show more features in common with primitive tetrapods than do the adult fishes. This is not so unusual, as the amphibians (like many other major higher groups of animals) probably evolved by retaining a number of juvenile features in their fish ancestor to bring about major changes in their adult morphology. This sort of evolutionary change is called heterochrony. Thus if an osteolepiform fish reaches sexual maturity earlier in successive generations, each generation will retain more

Comparison between the front limb (arm) skeletons of crossopterygian fishes and early amphibians. Note the striking similarity in the features of the humerus.

Eusthenopteron (fish) — humerus, ulna, radius

Acanthostega (amphibian) — humerus, ulna, radius

◀ A new reconstuction of *Acanthostega*, one of the earliest and most primitive fossil amphibians, from the Late Devonian of East Greenland. (After the work of Dr Mike Coates and Dr Jenny Clack)

▼ The foot of *Acanthostega* had eight digits.

juvenile characteristics and pass these on into maturity. The skulls of juvenile osteolepiforms thus look more like those of early tetrapods.

In reality the evolutionary mechanism is not so simple, as several processes are actually responsible for such major changes of form. The head, for instance, may look identical in fishes and early tetrapods. Here some juvenile characters of the fish are retained, and this is called paedomorphosis. However, the limb skeletons of the front and hind limbs have accelerated their development in the fish–tetrapod transition, thus additional growth stages have been added, forming more elongated limbs. This process is termed peramorphosis. Combinations of the two can rapidly bring about major changes in morphology without invoking great genetic distance, and this double-banger effect is termed dissoci-ated heterochrony. A wonderful parallel to the fish– tetrapod transition is seen in human evolution, where we retain the juvenile "flat-faced" and large-headed form of typical pongid apes (that is, paedomorphosis of the skull), yet our hind limbs have grown at faster, more accelerated rates (peramorphosis of the legs).

I can thus envisage the first amphibians evolving almost nonchalantly from advanced panderichthyid fishes, which were either sexually maturing at earlier generations (at younger stages of development) or growing more slowly and retaining more juvenile features into maturity, for some unexplained environmentally or diet-driven reason. These early tetrapods continued in their piscine lifestyle unperturbed by the fact that they actually modified the ends of their fins to form primitive fingers and toes. Perhaps in their early phases they may have possibly had short fin rays emerging to support the web between the digits. The expanded digits on the limbs may have become useful adaptations for changing direction more efficiently when manoeuvring in the water or for climbing up muddy riverbanks. The occasional ventures onto land were then facilitated as the wrist and ankle joints evolved by simple modification of the limb structure. Almost serendipitously the limbs became suddenly more practical, and longer jaunts away from the water may have resulted as new food sources were discovered or simply because such journeys enabled the animals to seek out quieter, isolated pools of water for safer breeding.

The real innovation was yet to come, some time

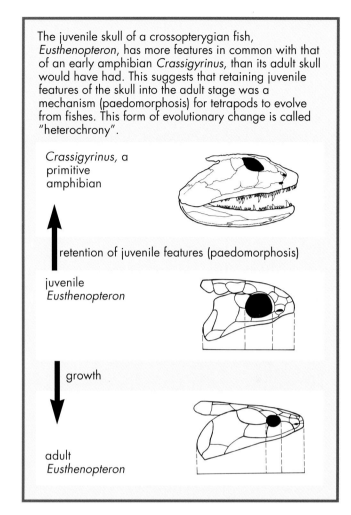

The juvenile skull of a crossopterygian fish, *Eusthenopteron*, has more features in common with that of an early amphibian *Crassigyrinus*, than its adult skull would have had. This suggests that retaining juvenile features of the skull into the adult stage was a mechanism (paedomorphosis) for tetrapods to evolve from fishes. This form of evolutionary change is called "heterochrony".

Crassigyrinus, a primitive amphibian

retention of juvenile features (paedomorphosis)

juvenile *Eusthenopteron*

growth

adult *Eusthenopteron*

within the next 50 million years or so, when amphibians became more proficient at terrestrial locomotion. Their next phases of evolution centred on the limbs becoming adapted to lift the animal off the ground and bear the weight under gravity, as opposed to the buoyancy provided in water. The axial skeleton underwent much modification as the backbone became strengthened to support the arms and legs for efficient movement on land. But despite these adaptations and the ensuing great radiation of amphibian forms in the Carboniferous Period, it was water they always had to return to for living and breeding.

In my opinion, it was the evolution of the hard-shelled egg, or amniote egg (containing the embryonic amnion), that was the greatest single advancement in the evolution of vertebrates from fish to human. This enabled the early tetrapods to venture away from the seas, rivers and lakes, and to seek new inland habitats. Their ancestral ties with the water had been finally broken. The invasion of continents by vertebrates was then just beginning. The rest is history.

A CLASSIFICATION OF FISHES

A classification of the Orders of fishes. Although this is based on various papers by recent workers there is aways some contention about the placement of problematic groups, and some taxonomic levels imposed through cladistic classifications will have no adequately named ranking.

SUPERPHYLUM AGNATHA

SUBPHYLUM CRANIATA

Infraphylum Myxinoidea (Myxiniformes)
Infraphylum Vertebrata

CLASS PTERASPIDOMORPHI

SUBCLASS ARANDASPIDA

Order Arandaspidiformes

SUBCLASS HETEROSTRACI

Suborder Cyathaspidiformes
Suborder Amphiaspidida
Suborder Pteraspidiformes
Order Theolodontida
RANK INDET. MYOPTERYGII
Suborder Hyperoartia
Order Petromyzontiformes
Order Anaspida
Suborder Rhyncholepidida

CLASS GALEASPIDA

Order Galeaspidiformes
Order Polybranchiaspidida
Order Huananaspidiformes

CLASS PITURIASPIDA

CLASS OSTEOSTRACI

Suborder Cornuata
Order Cephalaspidida
Order Zenaspida
Order Kiaeraspidida
Order Benneviaspidida
Order Thyestiida

SUBPHYLUM GNATHOSTOMATA

CLASS CHONDRICHTHYES

SUBCLASS ELASMOBRANCHII

SUPERORDER CLADOSELACHIMORPHA
Order Cladodontiformes

Class Chondrichthyes continued
Order Coronodontia
Order Symmoriida
Order Eugeneodontida
Order Squatinactida
SUPERORDER EUSELACHII
Order Ctenacanthiformes
Order Xenacanthida
Order Galeomorpha
Suborder Heterodontoidea
Suborder Orectoloboidea
Suborder Lamnoidea
Suborder Carcharinoidea
Suborder Hexanchoidea
Order Squalomorpha
Suborder Squaloidea
Suborder Pristiophoroidea
Suborder Squatinoidea
Order Batoidea
Suborder Torpenoidea
Suborder Pristoidea
Suborder Rhinobatoidea
Suborder Rajoidea
Suborder Myliobatoidea

SUBCLASS HOLOCEPHALI

SUPERORDER PARASELACHIMORPHA
Order Iniopterygii
Order Orodontiformes
Order Copodontiformes
Order Petalodontiformes
Order Psammosteiformes
SUPERORDER HOLOCEPHALIMORPHA
Order Cochliodontiformes
Order Chondrenchelyformes
Order Chimaeriformes
Suborder Echinochimaeroidei
Suborder Chimaeroidei
Suborder Cochliodontoidei
Suborder Helodontoidei
Suborder Menaspoidei
Suborder Squalorajoidei
Suborder Myriacanthoidei

CLASS PLACODERMI

Order Stensioellida
Order Pseudopetalichthyida
Order Ptyctodontida

Class Placodermi continued
Order Rhenanida
Order Acanthothoraci
Order Petalichthyida
Order Arthrodira
Infraorder Phyllolepidi
Infraorder Wuttagoonaspidi
Infraorder Actinolepidi
Suborder Phylctaeniina
Suborder Brachythoraci
Order Antiarchi
Suborder Yunnanolepidoidei
Suborder Sinolepidoidei
Suborder Bothriolepidoidei
Suborder Asterolepidoidei

CLASS ACANTHODII

Order Ischnacanthiformes
Order Climatiiformes
Suborder Diplacanthoidei
Order Acanthodiformes

CLASS OSTEICHTHYES

SUBCLASS ACTINOPTERYGII

INFRACLASS CHONDROSTEI
Order Palaeonisciformes
Suborder Platysomoidei
Order Haplolepiformes
Order Dorypteriformes
Order Tarrasiiformes
Order Ptycholepiformes
Order Pholidopleuriformes
Order Luganoiformes
Order Redfieldiiformes
Order Perleidiformes
Order Peltopleuriformes
Order Phanerorhynchiformes
Order Saurichthyiformes
Order Polypteriformes
Order Acipenseriformes
Suborder Chondrosteoidei
Suborder Acipenseroidei
Suborder Polyodontoidei
INFRACLASS NEOPTERYGII
Order Lepisosteiformes
Order Semionotiformes
Order Pycnodontiformes
Order Macrosemiiformes

Class Osteichthyes continued

 Order Amiiformes (Halecomorphi)

 Order Pachycormiformes

 Order Aspidorhynchiformes

DIVISION TELEOSTEI

 Order Pholidophoriformes

 Order Leptolepiformes

 Order Ichthyodectiformes

SUBDIVISION
OSTEOGLOSSOMORPHA

 Suborder Osteoglossoidei

 Suborder Notopteroidei

 Suborder Mormyroidei

SUBDIVISION ELOPOMORPHA

 Order Elopiformes

 Suborder Indet.

 Suborder Elopoidei

 Suborder Albuloidei

 Suborder Pachyrhizodontoidei

 Order Anguilliformes

 Suborder Saccopharyngidae

 Order Notocanthiformes

SUBDIVISION CLUPEOMORPHA

 Order Ellimmichthyiformes

 Order Clupeiformes

 Suborder Denticipitoidei

 Suborder Clupeoidei

SUBDIVISION EUTELEOSTEI

 Order Salmoniformes

 Suborder Esocoidei

 Suborder Argentinoidei

 Suborder Osmeroidei

 Suborder Galaxoidei

 Suborder Salmonoidei

 SUPERORDER OSTARIOPHYSI

 Order Gonorhynchiformes

 Suborder Gonorhynchoidei

 Suborder Chanoidei

 Order Characiformes

 Order Cypriniformes

 Order Siluriformes

 SUPERORDER STENOPTERYGII

 Order Stomiiformes

 Suborder Scopelomorpha

 Order Aulopiformes

 Suborder Enchodontoidei

 Suborder Cimolichthyoidei

 Suborder Halecoidei

 Suborder Alepisauroidei

 Suborder Aulopoidei

 Suborder Ichthyotringoidei

 Order Myctophiformes

 SUPERORDER INDET.

 Order Pattersonichthyiformes

 Order Ctenothrissiformes

SUPERORDER
PARACANTHOPTERYGII

Class Osteichthyes continued

 Order Percopsiformes

 Suborder Sphenocephaloidei

 Suborder Aphredoderoidei

 Suborder Percopsoidei

 Order Batrachoidiformes

 Order Gobiesciformes

 Order Lophiformes

 Suborder Lophioidei

 Suborder Antennaroidei

 Suborder Ceratioidei

 Order Gadiformes

 Suborder Muraenolepidoidei

 Suborder Gadoidei

 Suborder Macouroidei

 Order Ophidiiformes

SUPERORDER ACANTHOPTERYGII

SERIES ATHERINOMORPHA

 Order Atheriniformes

 Suborder Exocoetoidei

 Order Cyprinidontiformes

 Suborder Cyprinodontoidea

 Suborder Atherinoidei

SERIES PERCOMORPHA

 Order Beryciformes

 Suborder stephanoberycoidei

 Suborder Polymixoidei

 Suborder Dinopterygoidei

 Suborder Berycoidei

 Order Zeiformes

 Order Lampriformes

 Suborder Lamproidei

 Suborder Veliferoidei

 Suborder Trachipteroidei

 Suborder Styleophorgoidei

 Order Gasterosteiformes

 Order Syngnathiformes

 Suborder aulostomoidei

 Suborder Syngnathgoidei

 Order Synbranchiformes

 Order Indostomiformes

 Order Pegasiformes

 Order Dactylopteridae

 Order Scorpaeniformes

 Suborder Scorpaenoidei

 Suborder Hexagrammoidei

 Suborder Platycephaloidei

 Order Cottoidei

 Order Perciformes

 Suborder Percoidei

 Suborder Mugiloidei

 Suborder Sphyraenoidei

 Suborder Polynemoidei

 Suborder Labroidei

 Suborder Zoarcoidei

Class Osteichthyes continued

 Suborder Trachinoidei

 Suborder Notothenoidei

 Suborder Blenniodei

 Suborder Icosteoidei

 Suborder Schindleroidei

 Suborder Ammodytoidei

 Suborder Callionymoidei

 Suborder Goboidei

 Suborder Kuroidei

 Suborder Acanthuroidei

 Suborder Scombroidei

 Suborder Stromateoidei

 Suborder Anabantoidei

 Suborder Luciocephaloidei

 Suborder Chanoidei

 Suborder Mastacembeloidei

 Order Pleuronectiformes

 Suborder Psettodoidei

 Suborder Pleuronectoidei

 Suborder Soleioidei

 Order Tetraodontiformes

 Suborder Balistoidei

 Suborder Tetraodontoidei

**SUBCLASS DIPNOI
(LUNGFISHES)**

 Suborder Uranolophina

 Suborder Dipnorhynchina

 Suborder Speonesydrionina

**SUBCLASS
CROSSOPTERYGII**

 Order Actinistia (coelacanths)

 Order Onychodontiformes

DIVISION RHIPIDISTIA

 Order Porolepiformes

 Order Rhizodontiformes

 Order Osteolepiformes

 Order Panderichthyida

[DIVISION TETRAPODA

(ALL TETRAPODS)]

NB: Current classification of the Sarcopterygian fishes (dipnoans and crossopterygians) is under revision. The simple outline presented here represents a traditional view.

Glossary

Words in brackets are the vernacular term derived from the formal scientific term, or alternatives in common usage. Geological time periods and their dates can be found on the chart in the Introduction.

Acanthodii (acanthodian) A class of jawed fishes having bony fin spines preceding all fins. They lived between the Silurian and Permian Periods, and belong to three main orders, Climatiiformes Ischnacanthiformes and Acanthodiformes.

Acanthodiformes (acanthodiform) One of the three main orders of acanthodian fishes, characterised by one dorsal fin, lack of shoulder armour, and with well-developed gill rakers for a filter-feeding lifestyle.

Acanthothoraci (acanthothoracid = palaeacanthaspid) One of the seven major groups of placoderm fishes, that thrived during the Early Devonian. They have heavily ossified armours and skulls, often with elaborate surface ornament on the bones. Example: *Brindabellaspis*.

Acrodin A dense dentinous tissue found on the tips of the teeth in all ray-finned fishes (actinopterygians), excluding a few primitive Devonian forms such as *Cheirolepis*.

Actinistia (actinistian, coelacanth) One of the major groups of lobe-finned fishes (crossopterygians) characterised by the tassel-finned tail and lower jaw with short dentary and long angular bone, among other features. Known popularly as coelacanths, the only living member is *Latimeria chalumnae*.

Actinopterygii (actinopterygian) One of the three major subclasses of bony fishes (osteichthyans), containing the vast majority of all living fishes today. Characterised principally by having fins supported mainly by lepidotrichia; also ganoine layers in scales. Examples: trout, goldfish, salmon.

Anaspida (anaspids) Group of elongated, jawless fishes lacking bony armour and paired fins that lived in the Silurian and Devonian Periods of Euramerica. Examples: *Birkenia, Jamoytius*.

Antiarchi (antiarch) One of the seven major orders of placoderm fishes, characterised by having a long trunk shield, short head shield with eyes and nostrils centrally placed, and pectoral fins modified as bony props. Examples: *Bothriolepis, Asterolepis*.

Arthrodira (arthrodire, euarthrodire) One of the seven major orders of placoderm fishes, characterised by having two pairs of upper jaw tooth plates (superognathals), and includes more than 60 per cent of all known placoderm species. Arthrodires (meaning jointed neck) generally have well-developed neck joints between the head shield and trunk shield. Includes the largest placoderms, the gigantic predatory dinichthyids. Examples: *Coccosteus, Eastmanosteus*.

Asterolepidoidei (Asterolepididae, asterolepid) One of the major groups of antiarch placoderm fishes, characterised by having a short, robust pectoral fin appendage that does not extend beyond the long trunk shield; short head shield with very large orbital fenestra (central hole in head shield for eyes and nostrils). Examples: *Asterolepis, Pterichthyodes*.

Basibranchial The large ventral bone or series of bones in the gill arches, present in all gnathostome fishes. Osteichthyans typically have small toothed bones covering the basibranchials.

Batoidei (batoids, rays) Rays flattened chondrichthyans with large wing-like pectoral fins. Example: *Myliobatis*.

Benthic Bottom-dwelling (living near the floor or bed of the sea, river or lake).

Bothriolepidoidei (Bothriolepididae, bothriolepid) One of the major suborders of antiarch placoderm fishes characterised by having a long, segmented pectoral fin appendage that generally extends beyond the trunk shield; large head shield with small orbital fenestra (central hole for eyes and nostrils), and often has the postpineal plate recessed within the nuchal. Example: *Bothriolepis*.

Camuropiscidae (camuropiscid) Family of arthrodires (placoderms) known only from the Late Devonian Gogo Formation, Western Australia. They were streamlined elongate fishes, with large eyes and durophagous gnathal plates, some evolving long tubular snouts (rostral plates). Examples: *Camuropiscis, Rolfosteus*.

Canowindridae (canowindrid) Primitive family of Devonian osteolepiform crossopterygians with broad, flat heads and small eyes found only in Australia and Antarctica. Examples: *Koharolepis, Canowindra*.

Carcharhiniformes One of the major orders of sharks, containing the whalers and black-tipped reef sharks. Example: *Carcharhinus*.

Ceratodontidae (ceratodontids) Family of tooth-plated lungfishes with much reduced skull-roof pattern, and having a continuous median fin (anal and dorsal fins merged together with caudal fin). Mesozoic to recent. Examples: *Ceratodus, Neoceratodus*.

Ceratohyal A large bone in the first gill arch (hyoid arch) of jawed fishes situated between the hypohyal and the hyomandibular (or epihyal) elements. The ceratohyal is especially large in lungfishes and some crossopterygians.

Cheirolepidae (cheirolepid) Middle–Late Devonian family of primitive ray-finned fishes (actinopterygians) having tiny scales and long jaws, and lacking many advanced features such as acrodin caps on teeth. Example: *Cheirolepis* (only known genus).

Chondrichthyes (chondrichthyan) Cartilaginous fishes lacking endochondral bone, having the skin covered with very small placoid scales, such as the sharks, rays and chimaerids (holocephalans).

Chondrostei (chondrostean) Group of primitive ray-finned fishes (actinopterygians) which have developed a largely cartilaginous skeleton, although they still retain dermal bone in the skull, scales and parts of the pectoral fin girdle. Once all primitive ray-finned fishes (for example palaeoniscoids) were included in this group, but now it is used strictly to include the paddlefishes and sturgeons.

Climatiiformes (climatiiform) Order of acanthodian fishes having external dermal bones around the shoulder girdle, often expressed as elaborately ornamented pinnal and lorical plates around the pectoral fin insertion. Examples: *Climatius, Parexus, Euthacanthus*.

Coccosteomorph (coccosteid) Group or grade of arthrodiran placoderms having well developed jaws with predatory cusps, broad head shields and trunk shields with long median dorsal plates, often with a posterior spine developed. Example: *Coccosteus*.

Cochliodontiformes (cochliodonts) Order of Palaeozoic chondrichthyans within the Holocephalomorpha, having crushing tooth-plates with only two or three teeth in each jaw. They are closely related to the chimaerids as seen by their presence of prepelvic tenaculae preceding the claspers of males. Examples: *Cochliodus, Erismacanthus*.

Conodont Group of Palaeozoic–Triassic protochordates characterised by a worm-like body with bony rods supporting the tail, and having a set of phosphatic jaw-like elements in the head region, probably used for food capture or filtering. The microsocopic phosphatic elements are commonly found in marine sediments and are widely used for dating rock sequences.

Cosmine A layer of shiny external bony tissue found on the dermal bones of some extinct sarcopterygian fishes. It has an outer enameloid layer over a dentine layer which houses a system of flask-shaped cavities open to the surface by pores and interconnected to each other by canals; believed to be part of a complex vascular system for the skin.

Crossopterygii (crossopterygian) Subclass of lobe-finned predatory osteichthyan fishes characterised by their kinetic skulls, the braincase being divided into ethmosphenoid and oticco-occipital divisions. In most forms the primitive members have cosmine-covered bones and scales. Examples: *Latimeria, Eusthenopteron, Rhizodus*.

Dentine Hard, dense tissue found in the teeth of tetrapods also found in the teeth, spines and dermal bones of many primitive fishes. There are many kinds of fossil dentine based on the arrangements of nutritive canals and orientation of fibres within the dentine layer.

Dermal bone Bone formed in the dermis of the skin; often has a surface ornament. Includes all the outside skull bones of a fish and many plate-like bones in skulls of tetrapods. Scales are just small dermal bones that cover the fish's body.

Dermosphenotic Bone in the skull of certain actinopterygian fishes which flanks the skull-roof and meets the cheek unit behind the eye. It carries the main lateral line canal down to the cheek.

Dinichthyidae (dinichthyids) Family of large arthrodiran placoderms that had pachyosteomorph trunk shields and large pointed cusps on the upper and lower jaw bones. Some forms had lost the spinal plate from the armour.

Includes the largest ever placoderms, such as *Dunkleosteus* and *Gorgonichthys*, believed to be up to 8 m long.

Dipnoi (dipnoan) Lungfishes. One of the three main subclasses of osteichthyans, early lungfishes had a mosaic of bones forming the skull roof not comparable to the skull-roof patterns of other osteichthyans. The dentition in most forms is of hard crushing tooth-plates or a fine shagreen of denticles covering the palate, lower jaw bones and on bones resting on the ventral gill arches. Examples: *Dipterus*, *Neoceratodus*.

Dipteridae (dipterid) Family of Devonian lungfishes with well developed tooth-plates having many rows of teeth, two dorsal fins present, and with cosmine retained in part of the dermal skeleton. Example: *Dipterus*.

Distal Pertaining to part of an organism that is furthest away from the head (for example the hand is at the distal end of the arm).

Dorsal Top (the dorsal view of a car is that looking down on its roof). The dorsal surface of a bone is the surface seen from above when in life position.

Durophagous Feeding by crushing prey, particularly hard-shelled items. Such creatures usually have hardened tooth-plates rather than sharp pointed teeth.

Edestoidei (edestid) Group of fossil sharks that thrived in the Late Palaeozoic (especially in the Permian) characterised by their well-developed symphysial tooth whorls formed of laterally-compressed cutting blade-like teeth. Little is known of their skulls or body shape. Examples: *Helicoprion*, *Edestus*, *Agassizodus*.

Endochondral bone Bone formed from a cartilage core; replaces the cartilage frame, for example the limb bones.

Eusthenopteridae Family of advanced osteolepiform fishes which have lost the cosmine layer from their bones and scales, have deep, narrow heads, and most have lost the extratemporal bone from the skull. The scales are rounded with a boss on the basal surface. Examples: *Eusthenopteron*, *Marsdenichthys*, *Eusthenodon*.

Frontals Paired bones in the midline of the skull-roof in osteichthyan fishes and tetrapods. The frontals enclose the pineal opening or foramen in most fishes and tetrapods.

Galeaspida (galeaspid) Class of armoured jawless fishes that lived in China (south and north China terranes) during the Silurian to Late Devonian. They are characterised by a single-piece head shield with large median opening in front of the eyes and pineal opening.

Genus (pl. genera) Taxonomic term, a group of species which have characteristic features that unite them as a distinct group. e.g. *Homo* is the genus of man, species are *Homo sapeins*, *Homo erectus* and *Homo habilis*.

Gill arches Series of bones that support the gills of fishes, they articulate dorsally against the braincase, and ventrally they meet up (these are the ventral gill arch bones). The first arch is called the hyoid arch (hyomandibular, epihyal, ceratohyal, hypohyal, basihyal, etc.). The second and subsequent arches have no special names.

Head shield In placoderm fishes the series of bones that forms around the head. Includes the skull-roof unit and cheek units. May also apply to bony plates around the head of extinct jawless fishes.

Heterochrony Changing of timing or development of events, relative to the same events in the ancestor. Can explain morphological changes in the evolution of a lineage of species by simple changes in growth or development. Types of heterochronic changes include paedomorphosis (retention of sub-adult traits in the descendent adult) and peramorphosis (development of traits beyond the ancestral adult, extra growth stages added on).

Hexanchiformes (Hexanchidae etc.) Order (and family) of primitive neoselachian sharks having only one dorsal fin, commonly called seven-gilled sharks or cowsharks. The dentition of the upper and lower jaws is markedly different. Examples: *Hexanchus*, *Notorhynchus*.

Holocephali (holocephalan) Subclass of chondrichthyans having an opercular covering the gills, braincase fused to upper jaws, and crushing toothplates forming the dentition. Examples: *Chimaera*, *Callorhynchus*, *Edaphodon*.

Holocephalomorpha Grouping of chondrichthyan fishes including the modern Holocephali as well as several extinct related groups such as the petalodontid, cochliodont and iniopterygian chondrichthyans.

Holoptychiidae (holoptychiid) Family of advanced porolepiform crossopterygian fishes that have lost cosmine, and have rounded scales and a nariodal bone present in the snout region. Examples: *Glyptolepis*, *Holoptychius*.

Holostei (holostean) Term once used to group ray-finned fishes intermediate between the primitive palaeoniscoids and the advanced teleosts (for example gars and bowfins as well as many fossil forms; Ginglymodi, Halecomorphi). Recent classifications no longer recognise the group as monophyletic and place them as basal to all neopterygians.

Hybodontoidei (Hybodontidae, hybodontiform etc.) Sharks of a particular monophyletic lineage having stout dorsal fin spines and teeth with many cusps on each root, as well as having large tooth-like cranial scales. The group originated in the Early Carboniferous and became extinct at the close of the Cretaceous. Examples: *Hamiltonichthys*, *Hybodus*.

Hyomandibular Large bone of the hyoid arch (first gill arch) that braces the jaw articulation in many primitive fish groups, lying against the palatoquadrate. The hyomandibular has two articulations to the braincase in crossopterygians and

tetrapods, and in the latter becomes modified as the stapes bone of the inner ear.

Iniopterygii Group of chondrichthyan fishes allied to the holocephalans; they lived in the Late Palaeozoic of North America. The pectoral fins are high on the body and large with robust spines; the jaws may or may not be fused to braincase and separate teeth or tooth-plates can be present.

Incisoscutoidei (Incisoscutidae) Placoderm fishes of the Arthrodira, which have an incised trunk shield for the pectoral fin, durophagous dentition, and large orbits. Only known from the Late Devonian Gogo Formation, Western Australia, by two genera, *Incisoscutum* and *Gogosteus*.

Intracranial joint Joint between the two divisions of the braincase in crossopterygian fishes.

Jugal Bone in the cheek of osteichthyan fishes carrying the junction of the infraorbital and preopercular branches of the sensory-line canal.

Lachrymal Small bone in osteichthyan fishes that is the anterior most bone of the cheek unit. It sits under the eye and carries the infraorbital sensory-line canal.

Lamniformes (lamnid) Order (and family) of sharks having no nictitating eyelid. Includes the mako, porbeagle and great white sharks: *Isurus*, *Carcharodon*.

Lateral Side. The externally-facing side of an object or organism.

Lateral line Main sensory line canal system that runs along each side of a fish and onto the skull-roof.

Lepidotrichia Fin rays; bony rods that support the fins in osteichthyan fishes.

Lophosteiformes Order of primitive osteichthyan fishes closely allied to the first ray-finned fishes. They lived in the Late Silurian and are known only from scales, teeth and fragments of bone. Although close to early palaeoniscoid bone, they lack essential features like acrodin on teeth and peg and socket articulation on scales. Examples: *Lophosteus*, *Andreolepis*.

Maxilla (maxillary) Upper toothed jaw bone in fishes and tetrapods.

Megalichthyidae (megalichthyinids) Family of osteolepiform (crossopterygian) fishes having flat, rounded skulls with posterior tectal bone half enclosing the nostril; no pineal opening, and nasal bones deeply notched into the frontals. They retain cosmine over the bones and scales. Example: *Megalichthys*.

Monophyletic Any group of organisms that are united by sharing a unique set of features that evolved once within the group. This means that the group—whether a genus, family or higher taxonomic group—can be readily characterised by its specialised anatomical features. For example acanthodians are monophyletic because only they have paired spines preceding all the fins.

Naris (nares) External openings for the incurrent and excurrent nostrils of fishes.

Nasal Bone in the snout of osteichthyan fishes, it flanks the median snout bones and is notched for the nares.

Nasohypophysial opening Single opening on top of the head in osteostracans and lampreys which opens internally to the nasal capsules and the hypophysios.

Neopterygii Group of advanced actinopterygian fishes having ray-fins equal in number to their supports in the dorsal and anal fins, consolidated upper pharyngeal dentition, and a reduced or lost clavicle. Includes gars (*Lepisosteus*), bowfins (*Amia*) and teleosts.

Neoselachii Group of advanced chondrichthyan fishes having multilayered enameloid forming the crowns of the teeth, and root canals that penetrate the base of each tooth. Includes all living sharks and many fossil forms.

Neoteleostei Group of advanced actinopterygian fishes having, among other features, special dorsal constrictor muscle that works the upper pharyngeal jaw bones.

Neurocranium Braincase. In fishes this is the cartilaginous or bony block encasing the brain; it is generally separate from the dermal bones of the skull roof. The braincase in most fishes is either perichondral or endochondral.

Occipital Pertaining to the neck region or back of the skull.

Onychodontiformes (onychodontid) Order (and family) of crossopterygian fishes that lived only in the Devonian Period. They had whorls of curved teeth at the front of the lower jaw, and the cheek was simple, having only six bones. Examples: *Onychodus*, *Strunius*.

Operculogular bones Series of bones covering the gill arches in osteichthyan fishes. Includes the operculum and suboperculum, brachiostegal rays and gular plates.

Orbit Hole in the skull for the eye. Orbital notch may refer to part of the skull or cheek margin for the eye.

Orbital fenestra Centrally located hole in the head shield of some placoderms (antiarchs) for the eyes and nostrils.

Orodontidae Family of Palaeozoic sharks that had spineless elongated bodies with relatively small fins, and numerous rows of teeth that were broad and relatively blunt, for crushing food. Example: *Orodus*.

Osteolepiformes (osteolepiform) Order of crossopterygian fishes having a single external pair of nostrils and true internal nostril (choana), cheek with seven bones, including bar-like preopercular, and with basal scutes on fins. Includes *Osteolepis*, *Eusthenopteron*, and technically also includes the newest order, Panderichthyida.

Osteolepididae (osteolepid) Primitive family of osteolepiform fishes that is probably not a monophyletic group. They have cosmine on the bones and scales of most species (although lacking in *Glyptopomus*) and have rhombic scales. The skull has the extratemporal bone present. Examples: *Osteolepis, Thursius, Gyroptychius*.

Otolith Ear-stone in fishes; There may be up to three otoliths within each cavity of the inner ear, they act under gravity to orient the fish with respect to direction and speed.

Paedomorphosis The retention of sub-adult ancestral traits in the descendent adult (a process giving rise to new species in a lineage).

Palaeacanthaspida See **Acanthothoraci**

Palaeontology The study of fossils.

Palaeobiology The study of fossils as once-living orgasnisms, involves reconstructing their biological processes. (Also Palaeobotany is the study of fossil plants; Palaeozoology is the study of fossil animals.)

Palaeogeography The study of past continental positions and past environments, using largely information from rocks, fossils and magnetic data entrapped in minerals when rocks formed.

Palaeoniscoidei (palaeoniscid) A mixed grouping of primitive ray-finned fishes, such as most Devonian–Permian forms. This is not a monophyletic group and the term is used loosely to include early fossil actinopterygians having thick rhombic scales and long gapes that do not fit readily into one of the recognised higher monophyletic groups.

Panderichthyida (panderichthyid) Order (and family) of advanced osteolepiform fishes sharing more features in common with tetrapods than any other fishes. They are large-headed and long-bodied fish, lack dorsal and anal fins, have brow ridges on the skull and have ventrally facing nostrils near the mouth margin. Examples: *Panderichthys, Elpistostege*. Late Devonian of Euramerica.

Parallel evolution (convergence) Evolution of similar characters separately in two or more lineages of common ancestry.

Paraphyletic group A set of organisms which are not monophyletic, or not having the same common ancestry.

Parietals Paired bones in the rear mid-line of the skull-roof in fishes and tetrapods.

Parietal shield The set of bones covering the division of the skull-roof that is posterior to the intracranial joint in crossopterygians.

Peramorphosis Development of traits beyond that of the ancestral adult (a process giving rise to new species).

Percomorpha Group of advanced teleostean fishes within the Acanthopterygyii. It is an ill-defined group which may well be paraphyletic. Includes the orders Beryciformes, Lampridiformes, Zeiformes, Gasterosteiformes, Dactylopteriformes, Perciformes, Scorpaeniformes, Channiformes, Synbranchiformes, Tetra-odontiformes and Pleuronectiformes.

Perichondral bone Thin laminar bone precipitated around the soft tissues that pass within a cartilage block (for example the braincase in placoderms).

Petalichthyida (petalichthyid) Order of placoderms characterised by an elongate, flat skull having two pairs of paranuchals, long nuchal that reaches the pineal or preorbitals, and dorsally facing orbits. The lateral-line canals are situated deep in the dermal bones and open externally by pores or slits. The trunk shield lacks posterior laterals. Example: *Lunaspis*.

Petrodentine Hypermineralised dense tissue that forms the centres of tooth cusps in lungfish tooth-plates.

Phaneropleuridae (phaneropleurids) Family of Late Devonian tooth-plated lungfish typified by *Phaneropleuron* and *Scaumenacia*. The first dorsal fin is either very elongate and low or lost; the second dorsal fin is very large; the cheek is reduced.

Pholidophoridiformes One of the stem groups of teleosts that may not constitute a natural (monophyletic) grouping. *Pholidophorus* fits in somewhere between the Chondrostei and the neopterygians.

Phyllolepidi (phyllolepid) Late Devonian freshwater flattened placoderms having a large nuchal and median dorsal surrounded by a series of smaller marginal plates; the plates have a characteristic ornament of concentric radiating ridges. Example: *Phyllolepis*.

Phylogeny (phylogenetic) The evolution of a group; evolutionary relationships.

Pineal opening, foramen The so-called "third eye" in primitive vertebrates, a small opening in the skull-roof between the frontals (osteichthyans) or in the pineal plate of placoderms and agnathans. The pineal and parapineal organs are light-sensitive organs developed as protrusions from the brain. The pineal opening exists in many primitive tetrapods and may manifest itself in higher groups. René Descartes supposedly had a "pineal eye".

Placodermi (placoderm) A class of primitive shark-like fishes that lived from the Silurian to the end of the Devonian. They are characterised by their armour of overlapping dermal plates, usually forming a head shield and trunk shield.

Porolepidae Family of primitive porolepiform fishes having cosmine-covered bones and thick rhombic scales. Typified by the Early Devonian genus *Porolepis*,

the family is not characterised by any specialised features and could be a paraphyletic grouping.

Porolepiformes An order of Devonian crossopterygian fishes with broad, flat heads and small eyes; the cheek has a prespiracular plate, the lower jaws have tooth-whorls and the large teeth have a special dendront style of infolded enamel and dentine. Examples: *Holoptychius, Glyptolepis*.

Posterior The back end or surface of an organism or object (the posterior end of a car is the surface showing the rear number plate).

Premaxilla (premaxillary) Front upper paired toothed jaw bones in fishes and tetrapods which meet each other in the midline.

Proximal Pertaining to the part of an organism closest to the head (for example the proximal end of the arm is near the shoulder).

Pseudopetalichthyidae Primitive family of flattened placoderm fishes known only from the Early Devonian of Germany, represented by *Paraplesiobatis* and *Pseudopetalichthys*.

Ptyctodontida (ptyctodontids) Order of placoderm fishes with short trunk shields and reduced head shields; dentition of crushing tooth-plates; and long bodies with whip-like tails. The only placoderms in which males have claspers.

Quasipetalichthyidae (quasipetalichthyids) Family of peculiar petalichthyid placoderms known from fossils found in China. The head shield is broad with laterally-placed orbits, the nuchal plate is broad and elongate; the centrals are small and not touching the preorbital plates; pineal and endolymphatic openings are absent. Examples: *Quasipetalichthys, Eucaryaspis*.

Rhenanida Order of Early Devonian ray-like placoderm fishes. Example: *Gemuendina*.

Sarcopterygii (sarcopterygian) Grouping of crossopterygians and dipnoans as the "fleshy-finned fishes".

Sinolepidoidei (sinolepids) Suborder (and family) of antiarchan placoderm fishes having long occipital region on the head shields, with specialised trunk shields that have a very large ventral opening (which lacks the median ventral bone), and narrow ventral laminae on the anterior and posterior ventrolateral plates. Known only from China and Australia. Examples: *Sinolepis, Xichonolepis, Grenfellaspis*.

Stegotrachelidae Primitive family (probably not monophyletic) of mostly Devonian ray-finned fishes having rhombic scales, long cheek and single rostral bone and lacking accessory opercular bones. Examples: *Stegotrachelus, Mimia, Moythomasia*.

Stensioellidae Primitive family of placoderm fishes known only from the Early Devonian of Germany, represented only by *Stensioella*, which lacks the complete set of adjacent bones to form a rigid head-shield, although it does possess a ring of bones forming a trunk-shield.

Swim-bladder Internal gaseous exchange organ that keeps osteichthyan fish buoyant in the water. Evolved into the lung in lungfishes, some crossopterygians and tetrapods.

Symplectic bone Bone in the hyoid arch of some osteichthyan fishes that links the hyomandibular with the ceratohyal.

Synapomorphy A shared or derived character state in an organism. Only synapomorphies can define close relationships between taxonomic groups.

Teleostei (teleosteans) The largest group and most advanced of all ray-finned fishes, characterised principally by the presence of uroneural bones in the tail (elongated ural neural arches); unpaired basibranchial toothplates, a mobile premaxilla and internal carotid foramen within the parasphenoid bone. Includes most living fishes. Examples: salmon, trout, goldfish, tuna.

Tetrapod Four-legged vertebrate animal (includes all amphibians, reptiles, birds and mammals. Some of these forms may have secondarily lost limbs, such as snakes and legless lizards).

Thelodonti (thelodonts) Group of jawless fishes lacking bony armour but having a covering of thick dentine-covered scales. The isolated scales are all that have been found of most thelodonts. Silurian–Late Devonian. Examples: *Turinia, Loganiella, Phlebolepis*.

Trunk shield The series of bony plates that overlap to form a ring around the trunk of placoderm fishes, immediately behind the head. Includes median dorsal, anterior and posterior dorsolateral plates, anterior and posterior ventrolateral plates and may also include the spinal and median ventral plates. Articulates with the head shield in most placoderms.

Uroneural bones Found in the the tail of teleost fishes, these are elongated ural neural arches.

Ventral The underside or belly surface of an object (organism or bone).

Xenacanthidae (xenacanthids) Family of extinct sharks having teeth with two main cusps, and a well-developed median boss or button (lingual torus). Where the whole fish is known they have long bodies with large neck spines. Examples: *Xenacanthus, Pleuracanthus, Orthacanthus*.

Yunnanolepidoidei (yunnanolepids) Suborder of primitive antiarchan placoderms, known only from China and Vietnam, having poorly developed pectoral appendages without true brachial processes developed. Examples: *Yunnanolepis, Qujinolepis*.

Bibliography

The following references are a selected list of the mostly recent works that have been taken into account in the writing of the text and the creation of diagrams.

GENERAL

Carroll, R.L., 1987. *Vertebrate Palaeontology and Evolution*. Freeman and Co, New York. 698 pp.

Clarke, I.F. and Cook, B.J., 1990. *Perspectives of the Earth*. Australian Academy of Sciences. 2nd edn. 651 pp.

Frickhinger, K.A., 1991. *Fossilien Atlas Fische*. Mergus, Verlag für Natur-und Heimtierkunde, Melle, Germany. 1088 pp.

Long, J.A. (ed.), 1993. *Palaeozoic Vertebrate Biostratigraphy and Biogeography*. Belhaven Press, London. 368 pp.

Maisey, J.G., 1986. Heads and tails: a chordate phylogeny. *Cladistics* 2: 201–256.

Martin, A.P., Naylor, G.P.J. and S. R. Palumbi, 1992. Rates of mitochondrial DNA evolution in sharks are slow compared with mammals. *Nature* 357: 153–155.

McKinney, M.J. and McNamara, K.J., 1991. *Heterochrony. The evolution of ontogeny*. Plenum Press, New York, 437 pp.

Moy-Thomas, J.A. and Miles, R.S., 1971. *Palaeozoic Fishes*, 2nd edn. Chapman and Hall, London.

Obruchev, D.V., 1964. Agnatha and Fish. In Y.U. Orlov (ed.) *Fundamentals of Palaeontology* Vol. 11:1–522. Nauka, Moscow.

Pettijohn, E.J. 1975. *Sedimentary Rocks*. 3rd edn. Harper International Edition, Harper & Row, New York. 628 pp.

Skinner, B.J. and Porter, S.C., 1992. *The Dynamic Earth. An Introduction to Physical Geology*. 2nd edn. John Wiley and Sons, New York. 570 pp.

CHAPTER 1: ORIGINS OF VERTEBRATES

Aldridge, R. J., 1989. The soft body of evidence. *Natural History* 89/5: 6–11.

Blieck, A., 1992. At the origin of chordates. *Geobios* 25 (1): 101–113.

Briggs, D.E.G., Clarkson, E.N.K. & Aldridge, R.J., 1983. The conodont animal. *Lethaia* 16: 1–14.

Jeffries, R.P.S. and Lewis, D.N., 1978. The English Silurian fossil *Placocystites forbesianus* and the ancestry of the vertebrates. *Philosophical Transactions of the Royal Society of London* B 282: 205–323.

Long, J.A. and Burrett, C.F., 1989. Tubular phosphatic microproblematica from the Early Ordovician of China. *Lethaia* 22 (4): 439–446.

Maisey, J.G., 1988. Phylogeny of early vertebrate skeletal induction and ossification patterns. In *Evolutionary Biology*, M. Hecht (ed.), 22: 1–36.

Ritchie, A., 1985. *Ainiktozoan loganense* Scourfield, a protochordate? from the Silurian of Scotland. *Alcheringa* 9: 117–142.

Smith, M.M. and Hall, B. K. 1990. Development and evolutionary origins of vertebrate skeletogenic and odontogenic tissues. *Biological Reviews* 65: 277–373.

CHAPTER 2: AGNATHANS

Afanassieva, O., 1992. Some peculiarities of osteostracan ecology. In E. Mark-Kurik (ed.), *Fossil Fishes as Living Animals*. Academia 1, Academy of Sciences of Estonia, Tallinn, 1992, pp 61–69.

Arsenault, M. and Janvier, P., 1991. The anaspid–like craniates of the Escuminac Formation (Upper Devonian) from Miguasha (Québec, Canada), with remarks on anaspid–petromyzontid relationships. In Chang M.M., Liu Y.H. and Zhang G.R. (eds.), *Early Vertebrates and Related Problems of Evolutionary Biology*, Science Press, Beijing, pp.19–40.

Blieck, A., 1984. Les Hétérostracés Ptéraspidiformes, Agnathes du Silurien–Dévonien du continent Nord-Atlantique et des blocs avoisinants. *Cahiers de Paléontologie (Vertébrés)*, CNRS, Paris. 199 pp.

Blieck, A. and Janvier, P., 1991. Silurian Vertebrates. In M.G. Bassett et al. (eds), The Murchison Symposium (International Symposium on the Silurian System, Keele, 1989). *Palaeontology, Special Papers* 44: 345–389.

Denison, R.H., 1964. The Cyathaspididae, a family of Silurian and Devonian jawless vertebrates. *Fieldiana (Geology)* 13: 309–473.

Dineley, D.L., 1988. The radiation and dispersal of Agnatha in Early Devonian time. In N.J. McMillan, A.F. Embry and D.J. Glass (eds), *Devonian of the World* (Proceedings 2nd International Symposium on the Devonian System, Calgary, 1987). *Canadian Society of Petroleum Geologists Memoir* 14, III: 567–577.

Dineley, D.L. and Loeffler E.J., 1976. Ostracoderm faunas of the Delorme and associated Siluro–Devonian Formations, North Territories, Canada. *Palaeontology, Special Papers* 18: 1–214.

Elliott, D., 1987. A reassessment of *Astraspis desiderata*, the oldest North American vertebrate. *Science* 237: 190–192.

Forey, P.L., 1984. Yet more reflection on agnathan–gnathostome relationships. *Journal of Vertebrate Paleontology* 4: 330–343.

Forey, P. and Janvier, P., 1993. Agnathans and the origin of jawed vertebrates. *Nature* 361: 129–134.

Janvier, P., 1975. Les yeux des cyclostomes fossiles et le problème de l'origine des Myxinoides. *Acta Zoologica* 56: 1–9.

Janvier, P., 1984. The relationships of the Osteostraci and the Galeaspida. *Journal of Vertebrate Paleontology* 4: 344–358.

Janvier, P., 1985. Les Céphalaspides du Spitsberg : anatomie, phylogénie et systématique des Ostéostracés siluro–dévoniens; révision des Ostéostracés de la Formation de Wood Bay (Dévonien inférieur du Spitsberg). *Cahiers de Paléontologie*, CNRS, Paris. 244 pp.

Lund, R. and Janvier, P., 1986. A second lamprey from the Lower Carboniferous (Namurian) of Bear Gulch, Montana (USA). *Geobios* 19: 647–652.

Novitskaya, L.I., 1986. [Fossil agnathans of USSR – Heterostracans: cyathaspids, amphiaspids, pteraspids]. *Akademia Nauk SSSR, Trudy Paleontologicheskogo Instituta*, 219:1–159, Nauka, Moscow. (In Russian).

Novitskaya, L.I., 1992. Heterostracans: their ecology, internal structure and ontogeny. In E. Mark-Kurik (ed.), *Fossil Fishes as Living Animals*. Academia 1, Academy of Sciences of Estonia, Tallinn, 1992, pp. 51–61.

Obruchev, D.V., and Mark-Kurik, E., 1965. [Devonian Psammosteids (Agnatha, Psammosteidae) of the USSR]. *Eesti NSV Teaduste Akadeemia Geoloogia Institut*, 304 pp.

Pan Jiang and Chen Liezu, 1993. Geraspididae, a new family of Polybranchiaspidida (Agnatha) from Silurian of Northern Anhui. *Vertebrata Palasiatica* 31: 225–230.

Ritchie, A., 1964. New lights on the Norwegian Anaspida. *Norske Videnskaps Akademiens Skrifter (Matematiske–naturvidenskapslige Klasse)* 14: 1–35.

Ritchie, A., 1967. *Ateleaspis tessellata* Traquair, a non-cornuate cephalaspid from the Upper Silurian of Scotland. *Zoological Journal of the Linnean Society* London, 47: 69–81.

Ritchie, A., 1968a. New evidence on *Jamoytius kerwoodi*, an important ostracoderm from the Silurian of Lanarkshire, Scotland. *Palaeontology* 11: 21–39.

Ritchie, A., 1968b. *Phlebolepis elegans* Pander, an Upper Silurian thelodont of Oesel, with remarks on the morphology of thelodonts. In T. Ørvig (ed.), *Current Problems in Lower Vertebrate Phylogeny*, Nobel Symposium 4: 81–88, Almqvist and Wiksell, Stockholm.

Ritchie, A., 1980. The Late Silurian anaspid genus *Rhyncholepis* from Oesel, Estonia, and Ringerike, Norway. *American Museum Novitates* 2699: 1–18.

Ritchie, A., 1985. *Arandaspis prionotolepis*, the Southern four-eyed fish. In P. Rich and G. van Tets (eds), *Kadimakara: Extinct Vertebrates of Australia*, Pioneer Design Studios, Lilydale, pp. 95–106.

Ritchie, A. and Gilbert-Tomlinson J., 1977. First Ordovician vertebrates from the southern hemisphere. *Alcheringa* 1: 351–368.

Stensiö, E.A. 1927. The Downtonian and Devonian vertebrates of Spitsbergen. Part I. Family Cephalaspidae. *Skrifter om Svalbard og Ishavet* 12: 1–391.

Stensiö, E.A., 1932. *The Cephalaspids of Great Britain*. British Museum (Natural History), London.

Turner, S., 1973. Siluro–Devonian thelodonts from the Welsh Borderland. *Journal of the Geological Society of London* 129: 557–584.

Turner, S., 1991a. Monophyly and interrelationships of the Thelodonti. In Chang M.M., Liu Y.H. and Zhang G.R. (eds), *Early Vertebrates and Related Problems of Evolutionary Biology*, Science Press, Beijing, pp. 87–119.

Turner, S., 1991b. Palaeozoic Vertebrate Microfossils in Australasia. In P. Vickers-Rich, J.N. Monaghan, R.F. Baird and T.H. Rich (eds), *Vertebrate Palaeontology in Australasia*. Pioneer Design Studios with Monash University Publications Committee, Melbourne, pp. 42–464.

Turner, S., 1992. Thelodont lifestyles. In E. Mark-Kurik (ed.), *Fossil Fishes as Living Animals*. Academia 1, Academy of Sciences of Estonia, Tallinn 1992, pp. 21–40.

Turner, S. and Dring, R.S., 1981. Late Devonian thelodonts (Agnatha) from the Gneudna Formation, Carnarvon Basin, Western Australia. *Alcheringa* 5: 39–48.

Turner, S. and Young, G.C. 1992. Thelodont scales from the Middle–Late Devonian Aztec Siltstone, southern Victoria Land, Antarctica. *Antarctic Science* 4: 89–109.

Wang N.Z., 1991. Two new Silurian galeaspids (jawless craniates) from Zhejiang Province, China, with a discussion of galeaspid–gnathostome relationships. In Chang M.M., Liu Y.H., and Zhang G.R. (eds.), *Early Vertebrates and Related Problems of Evolutionary Biology*, Science Press, Beijing, pp. 41–65.

Wang N., and Dong Z., 1989. Discovery of late Silurian microfossils of Agnatha and fishes from Yunnan, China. *Acta Palaeontologica Sinica* 8:192–206.

Wang S., and Lan C., 1984. New discovery of polybranchiaspids from Yiliang County, Northeastern Yunnan. *Bulletin of the Geological Institute, Chinese Academy of Sciences* 9: 113–123.

Wang S., Dong Z. and Turner, S., 1986. Middle Devonian Turinidae (Thelodont, Agnatha) from Western Yunnan, China. *Alcheringa* 10: 315–325.

Wilson, M.V.W. and Caldwell, M.W., 1993. New Silurian and Devonian fork-tailed 'thelodonts' are jawless vertebrates with stomachs and deep bodies. *Nature* 361: 442–444.

Young, G.C., 1991. The first armoured agnathan vertebrates from the Devonian of Australia. In Chang M.M., Liu Y.H. & Zhang G.R. (eds), *Early Vertebrates and Related Problems of Evolutionary Biology*, Science Press, Beijing, pp. 67–85.

CHAPTER 3: CHONDRICHTHYANS

Cappetta, H., 1987. *Chondrichthyes II. Handbook of Palaeoichthyology, Vol. 3B*. H-P. Schultze, (ed.), Gustav Fischer Verlag, Stuttgart, New York. 193 pp.

Dick, J.R.F., 1978. On the Carboniferous shark *Tristychius arcuatus* Agassiz from Scotland. *Transactions of the Royal Society of Edinburgh* 70: 63–109.

Dick, J.F.R. and Maisey, J.G., 1980. The Scottish Lower Carboniferous shark *Onychoselache traquairi*. *Palaeontology* 23:363–374.

Duffin, C.J., 1988. The Upper Jurassic selachian *Palaeocarcharias* de Beaumont (1960). *Zoological Journal of the Linnean Society* 94: 271–286.

Duffin, C.J. and Ward, D.J., 1983. Neoselachian sharks teeth from the Lower Carboniferous of Britain and the Lower Permian of the USA. *Palaeontology* 26: 93–110.

Ginter, M. 1990. Late Famennian shark teeth from the Holy Cross Mountains, central Poland. *Acta Geologica Polonica* 40: 69–81.

Ginter, M. & Ivanov, A., 1992. Devonian phoebodont sharks teeth. *Acta Palaeontologica Polonica* 37: 55–75.

Goto, M., 1987. *Chlamydoselachus angineus*, a living cladodont shark. *Report of Japanese Group for Elasmobranch Studies* 23: 11–13.

Karatajute-Talimaa, V.N., 1973. *Elegestolepis grossi* gen. et sp. nov., ein neuer typ der Placoidschuppe aus dem oberen Silur der Tuwa. *Palaeontographica 143A*: 35–50.

Karatajute-Talimaa, V.N., Novitskaya, L.I., Rozman, Kh.S. and Sodov, Zh., 1990. *Mongolepis*, a new Lower Silurian genus of elasmobranch from Mongolia. *Palaeontological Journal* 1990, No. 1: 37–48.

Long, J.A., 1990. Late Devonian chondrichthyans and other microvertebrate remains from northern Thailand. *Journal of Vertebrate Paleontology* 10: 51–69.

Lund, R., 1974a. *Squatinactis caudispinatus*, a new elasmobranch from the Upper Mississippian of Montana. *Annals of the Carnegie Museum* 45: 43–55.

Lund, R., 1974b. *Stethacanthus altonensis* (Elasmobranchii) from the Bear Gulch Limestone of Montana. *Annals of the Carnegie Museum* 45: 161–178.

Lund, R., 1977a. A new petalodont (Chondrichthyes, Bradyodonti) from the Upper Mississippian of Montana. *Annals of the Carnegie Museum of Natural History* 46: 129–155.

Lund, R., 1977b. *Echinochimera meltoni*, new genus and species (Chimaeriformes) from the Mississippian of Montana. *Annals of the Carnegie Museum of Natural History* 46: 195–221.

Lund, R., 1982. *Harpagofututor volsellorhinus*, new genus and species (Chondrichthyes, Chondrenchelyiformes) from the Namurian Bear Gulch Limestone, *Chondrenchelys problematica* Traquair (Visean), and their sexual dimorphism. *Journal of Paleontology* 56: 938–958.

Lund, R., 1985a. Stethacanthid elasmobranch remains from the Bear Gulch Limestone (Namurian E2b) of Montana. *American Museum Novitates* 2828: 1–24.

Lund, R., 1985b. The morphology of *Falcatus falcatus* (St. John and Worthen), a Mississippian stethacanthid chondrichthyan from the Bear Gulch Limestone of Montana. *Journal of Vertebrate Paleontology* 5: 1–19.

Lund, R., 1986a. New Mississippian holocephalan (Chondrichthyes) and the evolution of the Holocephali. In *Teeth revisited: Proceedings of the VIIth International Symposium on dental morphology*, Paris 1986. Rusell, D.E., Santoro, J-P. and Sigogneau-Russell, D. (eds), *Memoires de la Museum Nationale, Histoire Naturelle de Paris (C)*, 53: 195–205.

Lund, R., 1986b. On *Damocles serratus* nov. gen. et sp. (Elasmobranchii:

Cladodontida) from the Upper Mississippian Bear Gulch Limestone of Montana. *Journal of Vertebrate Paleontology* 6: 12–19.

Lund, R., 1986c. The diversity and relationships of the Holocephali. In *Indo-Pacific Fish Biology: Proceedings of the Second International Conference on Indo-Pacific Fishes*, T. Uyeno, R. Arai, T. Taniuchi and K. Matsuura (eds), pp 97–106. Ichthyological Society of Japan.

Lund, R., 1989. New petalodonts (Chondrichthyes) from the Upper Mississipian Bear Gulch Limestone (Namurian E2b) of Montana. *Journal of Vertebrate Paleontology* 9: 350–368.

Lund, R., 1990. Chondrichthyan life history styles as revealed by the 320 million years old Mississipian of Montana. *Environmental Biology of Fishes* 27: 1–19.

Mader, H., 1986. Schuppen und Zahne von Acanthodiern und Elasmobranchiern aus dem Unter-Devon Spaniens (Pisces). *Gottinger Arbeiten zur Geologie und Paläontologie* 28: 1–59.

Maisey, J.G., 1977. The fossil selachian fishes *Palaeospinax* Egerton 1872, and *Nemacanthus* Agassiz 1837. *Zoological Journal of the Linnean Society* 60: 259–273.

Maisey, J.G., 1980. An evaluation of jaw suspension in sharks. *American Museum Novitates* 2706: 1–17.

Maisey, J.G., 1983. Cranial anatomy of *Hybodus basanus* Egerton from the Lower Cretaceous of England. *American Museum Novitates* 2758:1–64.

Maisey, J.G., 1984. Chondrichthyan phylogeny: a look at the evidence. *Journal of Vertebrate Paleontology* 4: 359–371.

Maisey, J.G., 1989a. *Hamiltonichthys mapesi* g. & sp. nov. (Chondrichthyes: Elasmobranchii) from the Upper Pennsylvanian of Kansas. *American Museum Novitates* 2931: 1–42.

Maisey, J.G., 1989b. Visceral skeleton and musculature of a Late Devonian shark. *Journal of Vertebrate Paleontology* 9: 174–190.

Moy-Thomas, J.A., 1936. The structure and affinities of the fossil elasmobranch fishes from the Lower Carboniferous rocks of Glencartholm, Eskdale. *Proceedings of the Zoological Society of London* 1936: 761–788.

Patterson, C., 1965. The phylogeny of the chimaeroids. *Philosophical Transactions of the Royal Society* B 249: 101–219.

Reif, W-E., 1978. Types of morphogenesis of the dermal skeleton in fossil sharks. *Paläontologische Zeitskrift* 52: 110–128.

St. John, O.H. and Worthen, A.H., 1875. Descriptions of fossil fishes. *Geological Survey of Illinois 1875,* 6: 245–488.

Turner, S., 1982. Middle Palaeozoic elasmobranchs remains from Australia. *Journal of Vertebrate Paleontology* 2: 117–131.

Turner, S., 1990. Lower Carboniferous shark remains from the Rockhampton district, Queensland. In S. Turner, R.A. Thulborn, and R. Molnar (eds), *De Vis Symposium volume*, *Memoirs of the Queensland Museum* 28: 65–73.

Turner, S. and Young, G.C., 1987. Shark teeth from the Early–Middle Devonian Cravens Peak Beds, Georgina Basin, Queensland. *Alcheringa* 11: 233–244.

Williams, M., 1985. The "cladodont" level sharks of the Pennsylvanian Black Shales of North America. *Palaeontographica* 190A: 83–158.

Williams, M., 1992. Jaws, the early years. Feeding behaviour in Cleveland Shale sharks. *Explorer,* Cleveland Museum of Natural History, Summer 1992: 4–8.

Young, G.C., 1982. Devonian sharks from south-eastern Australia and Antarctica. *Palaeontology* 25: 817–843.

Zangerl, R., 1981. *Paleozoic Chondrichthyes. Handbook of Paleoichthyology*. Vol 3A. H-P. Schultze, (ed.), Gustav Fischer Verlag, Stuttgart. 114 pp.

CHAPTER 4: ACANTHODIANS

Denison, R.H., 1979. *Acanthodii. Handbook of Paleoichthyology*. Pt 5. H-P. Schultze, (ed.), Gustav Fischer Verlag, Stuttgart. 62 pp.

Forey, P.L. and Young, V.T., 1985. Acanthodian and coelacanth fish from the Dinantian of Foulden, Berwickshire, Scotland. *Transactions of the Royal Society of Edinburgh, Earth Sciences* 76: 53–59.

Gross, W., 1971. Downtonische und Dittonische Acanthodier–Reste des Ostseegebietes. *Palaeontographica* 136A: 1–82.

Long, J.A., 1983. A new diplacanthoid acanthodian from the Late Devonian of Victoria. *Memoirs of the Association of Australasian Palaeontologists* 1: 51–65.

Long, J.A., 1986a. A new Late Devonian acanthodian fish from Mt Howitt, Victoria, Australia, with remarks on acanthodian biogeography. *Proceedings of the Royal Society of Victoria* 98: 1–17.

Long, J.A., 1986b. New ischnacanthid acanthodians from the Early Devonian of Australia, with a discussion of acanthodian interrelationships. *Zoological Journal of the Linnean Society* 87: 321–339.

Long, J.A., 1990. Chapter 11: Fishes. In K.J. McNamara (ed.), *Evolutionary Trends*, Belhaven Press, London, pp. 255–278.

Miles, R.S., 1966. The acanthodian fishes of the Devonian Plattenkalk of the Paffrath Trough in the Rhineland with an appendix containing a classification of the Acathodii and a revision of the genus *Homalacanthus*. *Arkiv für Zoologi* (Stockholm) ser. 2, 18: 147–194.

Miles, R.S., 1973a. Relationships of acanthodians. In P.H. Greenwood, R.S. Miles and C. Patterson (eds), *Interrelationships of Fishes*. Academic Press: London, pp. 63–104.

Miles, R.S., 1973b. Articulated acanthodian fishes from the Old Red Sandstone of England, with a review of the structure and evolution of the acanthodian shoulder girdle. *Bulletin of the British Museum of Natural History (Geology)* 24: 113–213.

Schultze, H-P., 1990. A new acanthodian from the Pennsylvanian of Utah, USA and the distribution of otoliths in vertebrates. *Journal of Vertebrate Paleontology* 10: 49–58.

Valiukevicius, J.J., 1985. Acanthodians from the Narva Regional Stage of the Main Devonian Field. Mosklas, Vilnius, 143 pp. [In Russian with English summary].

Watson, D.M.S., 1937. The acanthodian fishes. *Philosophical Transactions of the Royal Society of London* B 228: 49–146.

Woodward, A.S., 1906. On a Carboniferous fish fauna from the Mansfield district. *Memoirs of the National Museum of Victoria* 1: 1–32.

Young, G.C., 1989b. New occurrences of culmacanthid acanthodians (Pisces, Devonian) from Antarctica and southeastern Australia. *Proceedings of the Linnean Society of New South Wales* 111: 11–24.

Zidek, J., 1976. Kansas Hamilton Quarry (Upper Pennsylvanian) *Acanthodes*, with remarks on the previously reported North American occurrences of the genus. *The University of Kansas, Palaeontological Contributions Paper* 83: 1–41.

Zidek, J., 1980. *Acanthodes lundi*, new species (Acanthodii) and associated coprolites from uppermost Mississippian Heath Formation of central Montana. *Annals of the Carnegie Museum* 49: 49–78.

Zidek, J., 1985. Growth in *Acanthodes* (Acanthodii: Pisces), data and implications. *Paläontologisches Zeitschrift* 59: 147–166.

CHAPTER 5: PLACODERMS

Carr, R.K., 1991. Reanalysis of *Heintzichthys gouldii* (Newberry) an aspinothoracid arthrodire (Placodermi) from the Famennian of northern Ohio, with a review of brachythoracid systematics. *Zoological Journal of the Linnean Society* 103: 349–390.

Denison, R.H., 1978. *Placodermi. Handbook of Paleoichthyology*. Pt 2. H–P. Schultze, (ed.), Gustav Fischer Verlag, Stuttgart. 128 pp.

Dennis, K. and Miles, R.S., 1979. Eubrachythoracid arthrodires with tubular rostral plates from Gogo, Western Australia. *Zoological Journal of the Linnean Society* 67: 297–328.

Dennis, K. and Miles, R.S., 1981. A pachyosteomorph arthrodire from Gogo, Western Australia. *Zoological Journal of the Linnean Society* 73: 213–258.

Dennis–Bryan, K.D., 1987. A new species of eastmanosteid arthrodire (Pisces: Placodermi) from Gogo, Western Australia. *Zoological Journal of the Linnean Society* 90: 1–64.

Gardiner, B.G., 1984. The relationships of placoderms. *Journal of Vertebrate Paleontology* 4: 379–95.

Goujet, D., 1984. Les poissons placoderms du Spitsberg. Arthrodires Dolichothoraci de la Formation de Wood Bay (Dévonien inférieur). *Cahiers de Paleontologie, Sect. Vertébrés*, CNRS, Paris, 1–284.

Lelievre, H., Feist, R., Goujet, D. and Blieck, A., 1987. Les vertébrés Dévoniens de la Montagne Noire (Sud de la France) et leur apport à la phylogénie des pachyostéomorphes (placodermes arthrodires). *Paleovertebrata*, Montpellier 17 (1): 1–26.

Long, J.A., 1983. New bothriolepid fishes from the Late Devonian of Victoria, Australia. *Palaeontology* 26: 295–320.

Long, J. A., 1984. New phyllolepids from Victoria and the relationships of the group. *Proceedings of the Linnean Society of New South Wales* 107: 263–304.

Long, J.A., 1987. A new dinichthyid fish (Placodermi: Arthrodira) from the Upper Devonian of Western Australia, with a discussion of dinichthyid interrelationships. *Records of the Western Australian Museum* 13 (4): 515–540.

Long, J.A., 1988a. A new camuropiscid arthrodire (Pisces: Placodermi) from Gogo, Western Australia. *Zoological Journal of the Linnean Society* 94: 233–258.

Long, J.A., 1988b. Late Devonian fishes from the Gogo Formation, Western Australia. *National Geographic Research* 4: 436–450.

Long, J.A., 1990. Two new arthrodires (placoderm fishes) from the Upper Devonian Gogo Formation, Western Australia. *Memoirs of the Queensland Museum* 28: 51–63.

Long, J.A. and Werdelin, L., 1986. A new species of *Bothriolepis* (Placodermi, Antiarcha) from Tatong, Victoria, with descriptions of others from the state. *Alcheringa* 10: 355–399.

Long, J.A. and Young, G.C., 1988. Acanthothoracid remains from the Early Devonian of New South Wales, including a complete sclerotic capsule and pelvic girdle. *Memoirs of the Association of Australasian Palaeontologists* 7: 65–80.

Miles, R.S., 1967. Observations on the ptyctodont fish *Rhamphodopsis* Watson. *Journal of the Linnean Society of Zoology* 47: 99–120.

Miles, R.S., 1969. Features of placoderm diversification and the evolution of the arthrodire feeding mechanism. *Transactions of the Royal Society of Edinburgh* 68: 123–170.

Miles, R.S., 1971. The Holonematidae (placoderm fishes): a review based on new specimens of *Holonema* from the Upper Devonian of Western Australia. *Philosophical Transactions of the Royal Society of London* B263: 101–234.

Miles, R.S. and Dennis, K., 1979. A primitive eubrachythoracid arthrodire from Gogo, Western Australia. *Zoological Journal of the Linnean Society* 66: 31–62.

Miles, R.S. and Young, G.C., 1977. Placoderm interrelationships reconsidered in the light of new ptyctodontids from Gogo Western Australia. *Linnean Society Symposium Series* 4: 123–98.

Miles, R.S. and Westoll, T.S., 1968. The placoderm fish *Coccosteus cuspidatus* Miller ex Agassiz from the Middle Old Red Sandstone of Scotland. Part 1. Descriptive morphology. *Transactions of the Royal Society of Edinburgh* 67: 373–476.

Ritchie, A., 1973. *Wuttagoonaspis*, gen. nov., an unusual arthrodire from the Devonian of western New South Wales, Australia. *Palaeontographica* 143 A: 58–72.

Ritchie, A., 1975. *Groenlandaspis* in Antarctica, Australia and Europe. *Nature* 254: 569–573.

Ritchie, A., 1984. A new placoderm, *Placolepis* gen. nov. (Phyllolepidi) from the Late Devonian of New South Wales. *Proceedings of the Linnean Society of New South Wales* 107: 321–354.

Ritchie, A., Wang S., Young, G.C and Zhang G., 1992. The Sinolepidae, a family of antiarchs (placoderm fishes) from the Devonian of South China and eastern Australia. *Records of the Australian Museum* 44: 319–370.

Stensiö, E.A., 1963. Anatomical studies on the arthrodiran head. Part 1. Preface, geological and geographical distribution, the organization of the arthrodires, the anatomy of the head in the Dolichothoraci, Coccosteomorphi and Pachyosteomorphi. Taxonomic appendix. *Kungliga Svenska Vetenskapakadamiens Handlingar* 4 (9) 2: 1–419.

Wang S.T., 1987. A new antiarch from the Early Devonian of Guanxi. *Vertebrata Palasiatica* 25: 81–90.

Wang J.Q., 1982. New materials of Dinichthyidae. *Vertebrata Palasiatica* 20: 181–186.

Wang J.Q., 1991a. The Antiarch from the Early Silurian of Hunan. *Vertebrata Palasiatica* 29: 240–244.

Wang J.Q., 1991b. New material of *Hunanolepis* from the Middle Devonian of Hunan. In Chang M.M., Liu Y.H. and Zhang G.R. (eds.), *Early vertebrates and related problems of evolutionary biology*, Science Press, Beijing, pp. 213–247,

Wang J., 1991c. A fossil Arthrodira from Panxi, Yunnan. *Vertebrata Palasiatica* 29: 264–275.

White, E.I., 1978. The larger arthrodiran fishes from the area of the Burrinjuck Dam, NSW. *Transactions of the Zoological Society of London* 34: 149–262.

Young, G.C., 1979. New information on the structure and relationships of *Buchanosteus* (Placodermi: Euarthrodira) from the Early Devonian of New South Wales. *Zoological Journal of the Linnean Society* 66: 309–352.

Young, G.C., 1980. A new Early Devonian placoderm from New South Wales, Australia, with a discussion of placoderm phylogeny. *Palaeontographica,* 167A: 10–76.

Young, G.C., 1983. A new antiarchan fish (Placodermi) from the Late Devonian of southeastern Australia. *Bureau of Mineral Resources Journal of Australian Geology and Geophysics* 8: 71–81.

Young, G.C., 1984. Reconstruction of the jaws and braincase in the Devonian placoderm fish *Bothriolepis*. *Palaeontology* 27: 625–661.

Young, G.C., 1985. Further petalichthyid remains (placoderm fishes, Early Devonian) from southeastern Australia. *Bureau of Mineral Resources Australian Journal of Geology and Geophysics* 9:121–131.

Young, G.C., 1986. The relationships of placoderm fishes. *Zoological Journal of the Linnean Society* 88: 1–57.

Young, G.C., 1988. Antiarchs (placoderm fishes) from the Devonian Aztec Siltstone, southern Victoria Land, Antarctica. *Palaeontographica* A202: 1–125.

Young, G.C., 1990. New antiarchs (Devonian placoderm fishes) from Queensland, with comments on placoderm phylogeny and biogeography. *Memoirs of the Queensland Museum* 28: 35–50.

Young, G.C. and Gorter, J.D., 1981. A new fish fauna of Middle Devonian age from the Taemas-Wee Jasper region of New South Wales. *Bureau of Mineral Resources Bulletin* 209, (*Palaeontology Papers* 1981): 83–147.

Zhang G.R., 1978. The antiarchs from the Early Devonian of Yunnan. *Vertebrata Palasiatica* 16: 147–186.

Zhang G.R., 1984, New form of Antiarchi with primitive brachial process from Early Devonian of Yunnan. *Vertebrata Palasiatica* 22: 81–91.

Zhang G. and Liu Y.G., 1991. A new Antiarch from the Upper Devonian of Jianxi, China. In Chang M.M., Liu Y.H. and Zhang G.R. (eds.), *Early Vertebrates and Related Problems of Evolutionary Biology*, Science Press, Beijing, pp. 67–85.

CHAPTER 6: OSTEICHTHYANS

Chang M.M., 1982. The braincase of *Youngolepis*, a Lower Devonian crossopterygian from Yunnan, south–western China. *Papers in the Department of Geology University of Stockholm* 1 –113.

Chang M.M. and Smith, M.M., 1992. Is *Youngolepis* a porolepiform? *Journal of Vertebrate Paleontology* 12: 294–312.

Chang M.M. and Yu X.B., 1981. A new crossopterygian, *Youngolepis precursor*, gen. et sp. nov., from the Lower Devonian of E. Yunnan, China. *Scientia Sinica* 24: 89–97.

Chang M.M. and Yu X.B., 1984. Structure and phylogenetic significance of *Diabolichthys speratus* gen. et sp. nov., a new dipnoan–like form from the Lower Devonian of eastern Yunnan, China. *Proceedings of the Linnean Society of New South Wales* 107: 171–84.

Gross, W., 1968. Fraglich Actinopterygier Schuppen aus dem Silur Gotlands. *Lethaia* 1: 184–218.

Gross, W., 1971. *Lophosteus superbus* Pander: Zähne. Zahnknocken und besondere Schuppenformen. *Lethaia* 4: 131–152.

Janvier, P., 1978. On the oldest known teleostome fish *Andreolepis hedei* Gross (Ludlow of Gotland), and the systematic position of the lophosteids. *Esti NSV Teaduste Akadeemia Toimetised Koide Geologii* 27: 88–95.

Schaeffer, B., 1968. The origin and basic radiation of the Osteichthyes. *Nobel Symposium* 4 : 207–222.

Schultze, H–P., 1977. Ausgangform und Entwicklung der rhombischen Schuppen der Osteichthyes (Pisces). *Paläontologische Zeitskrift* 51: 152–68.

Smith M.M., 1979. SEM of the enamel layer in oral teeth of fossil and extant crossopterygian and dipnoan fishes. *Scanning Electron Microscopy* 2: 483–90.

Thomson, K.S., 1975. The biology of cosmine. *Bulletin of the Peabody Museum of Natural History* 40: 1–59.

Thomson, K.S., 1977. On the individual history of cosmine and a possible electroreceptive function of the pore–canal system in fossil fishes. In S.M. Andrews, R.S. Miles and A.D. Walker (eds.), *Problems in Vertebrate Evolution*, Academic Press, London, pp. 247–270.

Wiley, E.O., 1979. Ventral gill arch muscles and the interrelationships of gnathostomes, with a new classification of the Vertebrata. *Zoological Journal of the Linnean Society* 67: 149–79.

Zhang M. and Yu X., 1987. A nomen novum for *Diabolichthys* Chang et Yu 1981. *Vertebrata Palasiatica* 25: 79.

CHAPTER 7: ACTINOPTERYGIANS

Campbell, K.S.W. and Phuoc L.D., 1983. A Late Permian actinopterygian fish from Australia. *Palaeontology* 26: 33–70.

Dunkle, D. and Schaeffer, B., 1973. *Tegeolepis clarki* (Newberry), a palaeonisciform from the Upper Devonian Ohio Shale. *Palaeontographica 143A*: 151–8.

Gardiner, B.G., 1963. Certain palaeoniscoid fishes and the evolution of the snout in actinopterygians. *Bulletin of the British Museum of Natural History (Geology)* 8: 258–325.

Gardiner, B.G., 1967a. Further notes on palaeoniscoid fishes with a classification of the Chondrostei. *Bulletin of the British Museum of Natural History (Geology)* 14: 143–206.

Gardiner, B.G., 1967b. The significance of the preoperculum in actinopterygian evolution. *Zoological Journal of the Linnean Society* 47: 197–209.

Gardiner, B.G., 1973. Interrelationships of teleostomes. In P.H. Greenwood, R.S. Miles and C. Patterson (eds.), *Interrelationships of Fishes*, Academic Press, London, pp. 195–35.

Gardiner, B.G., 1984. Relationships of the palaeoniscoid fishes, a review based on new specimens of *Mimia* and *Moythomasia* from the Upper Devonian of Western Australia. *Bulletin of the British Museum of Natural History (Geology)* 37: 173–428.

Gardiner, B.G., 1986. Actinopterygian fish from the Dinantian of Foulden, Berwickshire, Scotland. *Transactions of the Royal Society of Edinburgh* 76: 61–6.

Gardiner, B.G. and Bartram, A.W.H., 1977. The homologies of ventral cranial fissures in osteichthyans. In S.M. Andrews, R.S. Miles and A.D. Walker (eds.), *Problems in Early Vertebrate Evolution*, Academic Press, London, pp. 227–245.

Grande, L., 1984. Palaeontology of the Green River Formation, with a review of the fish fauna. *Geological Survey of Wyoming, Bulletin* 63: 1–333.

Grande, L. and Cavender, T.M., 1991. Description and phylogenetic assessment of the monotypic Ostariostomidae (Teleostei). *Journal of Vertebrate Palaeontology* 11: 405–416.

Jessen, H., 1972. Schultergürtel und Pectoralflosse bei Actinopterygiern. *Fossils and Strata* 1: 1–101.

Kasantseva, A.A., 1981. Late Palaeozoic palaeoniscoids of East Khazakstan, systematics and phylogeny. *Trudy Palaeonotologiski Institut* 180: 1–139.

Kasantseva, A.A., 1982. Phylogeny of the lower actinopterygians. *Journal of Ichthyology* 22: 1–16.

Lauder, G.V. and Liem, K. F., 1983. The evolution and interrelationships of the actinopterygian fishes. *Bulletin of the Museum of Comparative Zoology* 150: 95–197.

Long, J.A., 1988. New palaeoniscoid fishes from the Late Devonian and Early Carboniferous of Victoria. *Memoirs of the Association of Australasian Palaeontologists* 7: 1–64.

Maisey, J.A. (ed.), 1990. *Santana Fossils*. Tropical Fish Hobbyist, New Jersey,

USA.

Nielsen, E., 1942. Studies on the Triassic fishes from East Greenland. 1. *Glaucolepis* and *Boreosomus*. *Meddelelser om Grønland* 138: 1–403.

Patterson, C., 1973. Interrelationships of holosteans. In P.H. Greenwood, R.S. Miles and C. Patterson (eds.), *Interrelationships of Fishes*, Academic Press, London, pp. 235–305.

Patterson, C., 1975. The braincase of pholidophoroid and leptolepid fishes, with a review of the actinopterygian braincase. *Philosophical Transactions of the Royal Society of London* B269: 275–579.

Patterson, C., 1982. Morphology and interrelationships of primitive actinopterygian fishes. *American Zoologist* 22: 241–59.

Pearson, D.M., 1982. Primitive bony fishes with especial reference to *Cheirolepis* and palaeonisciform actinopterygians. *Zoological Journal of the Linnean Society* 74: 35–67.

Pearson, D.M. and Westoll, T.S., 1979. The Devonian actinopterygian *Cheirolepis* Agassiz. *Transactions of the Royal Society of Edinburgh* 70: 337–99.

Poplin, C., 1974. Étude de quelques paléoniscidés pennsylvaniens du Kansas. *Cahiers de Paléontologie*, CNRS, Paris. 151 pp.

Rosen, D.E and Patterson, C., 1977. Review of ichthyodectiform and other Mesozoic teleost fishes and the theory and practise of classifying fossils. *Bulletin of the American Museum of Natural History* 158: 81–172.

Schaeffer, B., 1973. Interrelationships of chondrosteans. In P.H. Greenwood, R.S. Miles and C. Patterson (eds.), *Interrelationships of Fishes*, Academic Press, London, pp. 207–226.

Schaeffer, B. and Patterson, C., 1984. Jurassic fishes from the western United States, with comments on Jurassic fish distribution. *American Museum Novitates* 2796: 1–86.

Schultze, H-P., 1968. Palaeoniscoidea–Schuppen aus dem Unterdevon Australiens und Kanadas und aus Mittelsdevons Spitzbergens. *Bulletin of the British Museum of Natural History (Geology)* 16: 343–68.

Sorbini, L., 1983. La collezione baja di pesci e pianti fossili di Bolca. *Museo civico di Storia naturale Verona*. 117pp.

Tintori, A. and Sassi, D., 1992. *Thoracopterus* Bronn (Osteichthyes): a gliding fish from the Upper Triassic of Europe. *Journal of Vertebrate Palaeontology* 12: 265–283.

CHAPTER 8: DIPNOANS

Bemis, W.,1984. Paedomorphosis and the evolution of the Dipnoi. *Palaeobiology* 10: 293–307.

Bernacsek, G.M., 1977. A lungfish cranium from the Middle Devonian of the Yukon Territory, Canada. *Palaeontographica 157A*: 175–200.

Campbell, K.S.W., and Barwick, R.E., 1982a. A new species of the lungfish *Dipnorhynchus* from New South Wales. *Palaeontology* 25: 509–27.

Campbell, K.S.W. and Barwick, R.E., 1982b. The neurocranium of the primitive dipnoan *Dipnorhynchus sussmilchi* (Etheridge). *Journal of Vertebrate Paleontology* 2: 286–327.

Campbell, K.S.W. and Barwick, R.E., 1983. Early evolution of dipnoan dentitions and a new species *Speonesydrion*. *Memoirs of the Association of Australasian Palaeontologists* 1: 17–49.

Campbell, K.S.W. and Barwick, R.E.,1984. *Speonesydrion*, an Early Devonian dipnoan with primitive toothplates. *Palaeo Ichthyologica* 2: 1–48.

Campbell, K.S.W. and Barwick, R.E., 1987. Palaeozoic lungfishes, a review. *Journal of Morphology Supplement* 1: 93–131.

Campbell, K.S.W. and Barwick, R.E., 1988a. Geological and palaeontological information and phylogenetic hypotheses. *Geological Magazine* 125: 207–227.

Campbell, K.S.W. and Barwick, R.E., 1988b. *Uranolophus*: a reappraisal of a primitive dipnoan. *Memoirs of the Association of Australasian Palaeontologists* 7: 87–144.

Campbell, K.S.W. and Barwick, R.E., 1990. Palaeozoic dipnoan phylogeny: functional complexes and evolution without parsimony. *Paleobiology* 16: 143–169.

Campbell, K.S.W. and Barwick, R.E., 1991. Teeth and tooth plates in primitive lungfish and a new species of *Holodipterus*. In Chang M.M., Liu Y.H. and Zhang G.R. (eds.), *Early Vertebrates and Related Problems of Evolutionary Biology*, Science Press, Beijing, pp. 429–440.

Campbell, K.S.W. and Smith, M.M., 1987. The Devonian dipnoan *Holodipterus*: dental variation and remodelling growth mechanisms. *Records of the Australian Museum* 38: 131–67.

Denison, R.H., 1968. Early Devonian lungfishes from Wyoming, Utah and Idaho. *Fieldiana (Geology)* 17: 353–413.

Lehman, J.P., 1959. Les dipneustes du Dévonien supérieur du Groenland. *Meddelelser om Grønland* 160 (4): 1–58.

Long, J.A., 1992a. Cranial anatomy of two new Late Devonian lungfishes, from Mt. Howitt, Victoria. *Records of the Australian Museum* 44: 299–319.

Long, J.A., 1992b. Gogodipterus paddyensis (Miles) gen. nov., a new chirodipterid lungfish from the Late Devonian Gogo Formation, Western Australia. *The Beagle*, Northern Territory Museum 9: 11–21.

Long, J.A., 1993. Cranial ribs in Devonian lungfishes and the origin of dipnoan air-breathing. *Memoirs of the Association of Australasian Palaeontologists* 15: 199–209.

Long, J.A. and Campbell, K.S.W., 1985. A new lungfish from the Early Carboniferous of Victoria. *Proceedings of the Royal Society of Victoria* 97: 87–93.

Long, J.A., Campbell, K.S.W. and Barwick, R.E., 1994. A new lungfish from the Early Devonian of Victoria, Australia. *Journal of Vertebrate Palaeontology* 14 (1): 127–31.

Marshall, C.R., 1987. Lungfish: phylogeny and parsimony. *Journal of Morphology Supplement* 1: 151–62.

Miles, R.S., 1977. Dipnoan (lungfish) skulls from the Upper Devonian of Western Australia. *Zoological Journal of the Linnean Society* 61: 1–328.

Pridmore, P.A. and Barwick, R.E., 1993. Post-cranial morphologies of the Late Devonian dipnoans Griphognathus and Chirodipterus and locomotor implications. *Memoirs of the Association of Australasian Palaeontologists* 15: 161–82.

Pridmore, P.A., Campbell, K.S.W. and Barwick, R.E., 1994. Morphology and phylogenetic position of the holodipteran dipnoans of the Upper Devonian Gogo Formation of northwestern Australia. *Philosophical Transactions of the Royal Society of London,* B 344: 105–64.

Schultze, H–P., 1969. *Griphognathus* Gross, ein langschnauziger Dipnoer aus dem Oberdevon von Bergisch–Gladbach (Rheinisches Schiefergebirge). *Geologica et Palaeontologica* 3: 21–61.

Schultze, H–P., 1987. Dipnoans as sarcopterygians. *Journal of Morphology Supplement* 1: 39–74.

Schultze, H–P., 1992. A new long–headed dipnoan (Osteichthyes) from the Middle Devonian of Iowa, USA. *Journal of Vertebrate Palaeontology* 12: 42–58.

Schultze, H–P. and Campbell, K.S.W., 1987. Characterization of the Dipnoi, a monophyletic group. *Journal of Morphology Supplement* 1: 25–37.

Schultze, H–P. and Marshall, C.R., 1993. Contrasting the use of functional complexes and isolated charcaters in lungfish evolution. *Memoirs of the Association of Australasian Palaeontologists* 15: 211–224.

Smith, M.M., 1977. The microstructure of the dentition and dermal ornament of three dipnoans from the Devonian of Western Australia: a contribution towards dipnoan interrelations, and morphogenesis, growth and adaptation of skeletal tissues. *Philosophical Transactions of the Royal Society of London* B 281: 29–72.

Smith, M.M. and Campbell, K.S.W., 1987. Comparative morphology, histology and growth of dental plates of the Devonian dipnoan Chirodipterus. *Philosophical Transactions of the Royal Society of London* 317: 329–363.

Watson, D.M.S. and Gill, E.L., 1923. The structure of certain Palaeozoic Dipnoi. *Zoological Journal of the Linnean Society* 35: 163–216.

Westoll, T.S., 1949. On the evolution of the Dipnoi. In G.L. Jepson, G.G. Simpson & E. Mayr (eds.), *Genetics, Palaeontology and Evolution*, University Press, Princeton, pp. 121–184.

CHAPTER 9: CROSSOPTERYGIANS

Ahlberg, P.E., 1989. Paired fin skeletons and relationships of the fossil group Porolepiformes (Osteichthyes: Sarcopterygii). *Zoological Journal of the Linnean Society* 96: 119–166.

Ahlberg, P.E., 1991. A re-examination of sarcopterygian interrelationships, with special reference to the Porolepiformes. *Zoological Journal of the Linnean Society* 103: 241–287.

Ahlberg, P.E., 1992. The palaeoecology and evolutionary history of the porolepiform sarcopterygians. In E. Mark–Kurik (ed.), *Fossil Fishes as Living Animals*, Academy of Sciences of Estonia, Tallinn, pp.71–90.

Andrews, S.M., 1973. Interrelationships of crossopterygians. In P.H. Greenwood, R.S. Miles and C. Patterson (eds.), *Interrelationships of Fishes*, Academic Press, London, pp. 137–177.

Andrews, S.M., 1977. The axial skeleton of the coelacanth, *Latimeria*. In S.M. Andrews, R.S. Miles and A.D. Walker (eds.), *Problems in Vertebrate Evolution*, Academic Press, London, pp. 271–288.

Andrews, S.M., 1985. Rhizodont crossopterygian fish from the Dinantian of Foulden, Berwickshire, Scotland, with a re–evaluation of this group. *Transactions of the Royal Society of Edinburgh (Earth Sciences)* 76: 67–95.

Andrews, S.M. and Westoll, T.S., 1970a. The postcranial skeleton of *Eusthenopteron foordi* Whiteaves. *Transactions of the Royal Society of Edinburgh* 68: 207–329.

Andrews, S.M. and Westoll, T.S., 1970b. The postcranial skeleton of rhipidistian fishes excluding *Eusthenopteron*. *Transactions of the Royal Society of Edinburgh* 68: 391–489.

Cloutier, R., 1991a. Patterns, trends, and rates of evolution within the Actinistia. *Environmental Biology of Fishes* 32: 23–58.

Cloutier, R., 1991b. Interrelationships of Palaeozoic actinistians: patterns and trends. In Chang M.M., Liu Y.H. and Zhang G.R. (eds.). *Early Vertebrates and Related Problems of Evolutionary Biology*, Science Press, Beijing, pp. 379–428.

Cloutier, R., in press. Phylogenetic status, basal taxa, and interrelationships of lower sarcopterygian groups. *Zoological Journal of the Linnean Society* (in press).

Forey, P.L., 1980. *Latimeria*: a paradoxical fish. *Proceedings of the Royal Society of London* B 208 : 369–84.

Forey, P.L., 1981. The coelacanth *Rhabdoderma* in the Carboniferous of the British Isles. *Palaeontology* 24: 203–29.

Jarvik, E., 1944. On the dermal bones, sensory canals and pit–lines of the skull in *Eusthenopteron foordi* Whiteaves, with some remarks on *E. save–soderberghi* Jarvik. *Kungliga Svenska Vetenskapakadamiens Handlingar* 3 (21) 3: 1–48.

Jarvik, E., 1948. On the morphology and taxonomy of the Middle Devonian osteolepid fishes of Scotland. *Kungliga Svenska Vetenskapakadamiens Handlingar* 3 (25): 1–301.

Jarvik, E., 1950a. On some osteolepiform crossopterygians from the Upper Old Red Sandstone of Scotland. *Kungliga Svenska Vetenskapakadamiens Handlingar* 2: 1–35.

Jarvik, E., 1954. On the visceral skeleton in *Eusthenopteron* with a discussion of the parasphenoid and palatoquadrate in fishes. *Kungliga Svenska Vetenskapakadamiens Handlingar* (4) 5: 1–104.

Jarvik, E., 1972. Middle and Upper Devonian Porolepiformes from East Greenland with special reference to *Glyptolepis groenlandica* n.sp. *Meddelelser om Grønland* 187 (2): 1–295.

Jarvik, E., 1980. *Basic Structure and Evolution of Vertebrates*. Vols. 1, 2. Academic Press, London.

Jarvik, E., 1981. Review of "Lungfishes, tetrapods, palaeontology and plesiomorphy". *Systematic Zoology* 30: 378–84.

Jarvik, E., 1985. Devonian osteolepiform fishes from East Greenland. *Meddelelser om Grønland* 13: 1–52.

Jessen, H., 1966. Die Crossopterygier des Oberen Plattenkalkes (Devon) der Bergisch–Gladbach–Paffrather Mulde (Rheinisches Schiefergebirge) unter Berücksichtigung von amerikanischem und europäischem *Onychodus*–material. *Arkivs für Zoologi* 18: 305–89.

Jessen, H., 1967. The position of the Struniiformes (*Strunius* and *Onychodus*) among crossopterygians. *Colloques internationale de CNRS* 163: 173–80.

Jessen, H., 1973. Weitere Fishrestes aus dem Oberen Plattenkalk der Bergisch–Gladbach–Paffrather Mulde (Oberdevon, Rheinisches Schiefergebirge). *Palaeontographica* 143A: 159–87.

Jessen, H., 1980. Lower Devonian Porolepiformes from the Canadian Arctic with special reference to *Powichthys thorsteinssoni* Jessen. *Palaeontographica* 167 A: 180–214.

Long, J.A., 1985a. The structure and relationships of a new osteolepiform fish from the Late Devonian of Victoria, Australia. *Alcheringa* 9: 1–22.

Long, J.A., 1985b. A new osteolepidid fish from the Upper Devonian Gogo Formation, Western Australia. *Records of the Western Australian Museum* 12: 361–377.

Long, J.A., 1985c. New information on the head and shoulder girdle of *Canowindra grossi* Thomson, from the upper Devonian Mandagery Sandstone, New South Wales. *Records of the Australian Museum* 37: 91–99.

Long, J.A., 1987. An unusual osteolepiform fish from the Late Devonian of Victoria, Australia. *Palaeontology* 30: 839–852.

Long, J.A., 1989. A new rhizodontiform fish from the Early Carboniferous of Victoria, Australia, with remarks on the phylogentic position of the group. *Journal of Vertebrate Paleontology* 9: 1–17.

Long, J.A., 1991. Arthrodire predation by *Onychodus* (Pisces, Crossopterygii) from the Upper Devonian Gogo Formation, Western Australia. *Records of the Western Australian Museum* 15: 369–371.

Lund, R. and Lund, W., 1985. The coelacanths from the Bear Gulch Limestone (Namurian) of Montana and the evolution of the coelacanthiformes. *Bulletin of the Carnegie Museum of Natural History* 25: 1–74.

Schultze, H–P., 1973. Crossopterygier mit heterozerker Schwanzflosse aus dem Oberdevon Kanadas, nebst einer Beschreibung von Onychodontida–Resten aus dem Mitteldevon Spaniens und aus dem Karbon der USA. *Palaeontographica* 143 A: 188–208.

Schultze, H–P. and Arsenault, M., 1987. *Quebecius quebecius* (Whiteaves), a porolepiform crossopterygian (Pisces) from the Late Devonian of Quebec, Canada. *Canadian Journal of Earth Sciences* 24: 2351–61.

Stensiö, E.A., 1937. On the Devonian coelacanthids of Germany with special reference to the dermal skeleton. *Kungliga Svenska Vetenskapakadamiens Hanlingar* (3) 16 (4): 1–56.

Thomson, K.S., 1969. The biology of the lobe-finned fishes. *Biological Reviews* 44: 91–154.

Thomson, K.S., 1973. Observations on a new rhipidistian fish from the Upper Devonian of Australia. *Palaeontographica* 143 A: 209–20.

Vorobyeva, E., 1962. Rhizodont crossopterygians from the Devonian main field of the USSR. *Trudy Paleontological Institute* 104: 1–108.

Vorobyeva, E., 1975. Formenvielfalt und Verwandtschaftsbeiziehungen der osteolepidida (crossopterygier, Pisces). *Paläontologische Zeitskrift* 49: 44–54.

Vorobyeva, E., 1977. Morphology and nature of evolution of crossopterygian fish. *Trudy Paleontological Institute* 163: 1–239.

Vorobyeva, E., 1980. Observations on two rhipidistian fishes from the Upper

Devonian of Lode, Latvia. *Zoological Journal of the Linnean Society* 70: 191–201.

Young, G.C., Long, J.A. and Ritchie, A., 1992. Crossopterygian fishes from the Devonian of Antarctica: systematics, relationships and biogeographic significance. *Records of the Australian Museum Supplement* 14:1–77.

CHAPTER 10: THE GREAT STEP IN EVOLUTION

Ahlberg, P.E., 1991. Tetrapod or near tetrapod fossils from the Upper Devonian of Scotland. *Nature* 354: 298–301.

Campbell, K.S.W. and Bell, M.W., 1977. A primitive amphibian from the Late Devonian of New South Wales. *Alcheringa* 1: 369–382.

Chang M.M., 1991. "Rhipidistians", dipnoans and tetrapods. In H–P. Schultze and L.Trueb (eds.), *Origin of the Higher Groups of Tetrapods*, Comstock Publishing, Ithaca, pp 3–28.

Coates, M.I. and Clack, J.A., 1990. Polydactyly in the earliest known tetrapod limbs. *Nature* 347: 66–69.

Coates, M.I. and Clack, J.A., 1991. Fish–like gills and breathing in the earliest known tetrapod. *Nature* 352: 234–236.

Forey, P.L., Gardiner, B.G. and Patterson, C., 1991. The lungfish, the coelacanth, and the cow revisited. In H–P. Schultze and L.Trueb (eds.), *Origin of the Higher Groups of Tetrapods*, Comstock Publishing, Ithaca, pp 145–172.

Gardiner, B.G., 1980. Tetrapod ancestry: a reappraisal. In A.L. Panchen (ed.), *The terrestrial environment and the origin of land vertebrates*, Academic Press, London, pp 177–185.

Lebedev, O.A. and Clack, J.A., 1994. Devonian tetrapod remains from Andreyevka, Tula, Russia. *Palaeontology* 37.

Long, J.A., 1990. Heterochrony and the origin of tetrapods. *Lethaia* 23: 157–166.

Panchen, A.L., 1985. On the amphibian *Crassigyrinus scoticus* Watson from the Carboniferous of Scotland. *Philosophical Transactions of the Royal Society of London* B 309: 505–568.

Panchen, A.L., 1991. The early tetrapods: classification and the shapes of cladograms. In H–P. Schultze and L.Trueb (eds.), *Origin of the Higher Groups of Tetrapods*, Comstock Publishing, Ithaca pp. 110–144.

Panchen, A.L., and Smithson, T.R., 1987. Character diagnosis, fossils and the origin of the tetrapods. *Biological Reviews* 62: 341–438.

Rosen, D.E., Forey, P.L., Gardiner, B.G. and Patterson, C., 1981. Lungfishes, tetrapods, palaeontology and plesiomorphy. *Bulletin of the American Museum of Natural History* 167: 159–276.

Schultze, H–P., 1970. Folded teeth and the monophyletic origin of the tetrapods. *American Museum Novitates* 2408: 1–10.

Schultze, H–P., 1984. Juvenile specimens of *Eusthenopteron foordi* Whiteaves 1881 (osteolepiform rhipidistian, Pisces), from the Upper Devonian of Miguashua, Quebec, Canada. *Journal of Vertebrate Paleontology* 4: 1–16.

Schultze, H–P., 1991. A comparison of controversial hypoptheses on the origin of tetrapods. In H–P. Schultze and L.Trueb (eds.), *Origin of the Higher Groups of Tetrapods*, Comstock Publishing, Ithaca, pp 29–67.

Schultze, H–P. and Arsenault, M., 1985. The panderichthyid fish *Elpistostege*: a close relative of tetrapods? *Palaeontology* 28: 293–310.

Thomson, K.S., 1968. A critical review of the diphyletic theory of rhipidistian–amphibian relationships. *Nobel Symposium* 4: 285–305.

Vorobyeva, E. and Schultze, H–P., 1991. Description and systematics of panderichthyid fishes with comments on their relationships to tetrapods. In H–P. Schultze and L.Trueb (eds.), *Origin of the Higher Groups of Tetrapods*, Comstock Publishing, Ithaca, pp 68–109.

Vorobyeva, E. and Kuznetsov, A., 1992. The locomotor apparatus of *Panderichthys rhombolepis* (Gross), a supplement to the problem of fish–tetrapod transition In E. Mark–Kurik (ed.), *Fossil fishes as living animals*, Academy of Sciences of Estonia, Tallinn, pp 131– 140.

Warren, A.A., Jupp, R. and Bolton, B., 1986. Earliest tetrapod trackway. *Alcheringa* 10: 183–186.

Warren, J.W. and Wakefield, N.A., 1972. Trackways of tetrapod vertebrates from the Upper Devonian of Victoria, Australia. *Nature* 238: 469–470.

Index

MONGOLIA

NINGXIA

JAPAN

• Fenhsiang

CHINA

• Yunnan

VIETNAM

① Cobar
② Talbragar River
③ Somersby/Gosford
④ Jemalong Gap/Forbes
⑤ Canowindra
⑥ Grenfell
⑦ Brookvale
⑧ Taemas–Wee Jasper/Burrinjuck Dam
⑨ Cooma
⑩ Bunga Beds
⑪ Genoa River
⑫ Mansfield
⑬ Buchan
⑭ Bells Point/Waratah Bay
⑮ Koonwarra
⑯ Grampians
⑰ Mt Howitt

NORTHERN TERRITORY

• Gogo Station

QUEENSLAND

• Alice Springs   Toko Range

Shark Bay

A U S T R A L I A

• Kalbarri   **WESTERN AUSTRALIA**

Redbank Plains

SOUTH AUSTRALIA   Lightning Ridge •

• Ediacara   **NEW SOUTH WALES**   ①②

VICTORIA   ⑯ ⑮ ⑫⑰ ⑬⑪⑩ ⑭ ④⑤⑥ ③ ⑦ ⑧⑨

Sydney Basin

NORTHWEST TERRITORIES
(Delorme Formation)

BRITISH COLUMBIA
(Burgess Shale)

C A N A D A

MONTANA
• Bear Gulch

WYOMING
• Green River

NEBRASKA

NEVADA

COLORADO   KANSAS

U N I T E D   S T A T E S   O F

M A P   O F   F
including principal reg

Continued fr